To My Son...

The Life & War Remembrances of Captain Mordecai Myers, 13th United States Infantry 1812-1815

Including
Letters and Reminiscences
Pertaining to His Early Life
(1780-1814),
and Incidents in the
War of 1812

Edited,
with Biographical Introduction
by

Neil B. Yetwin

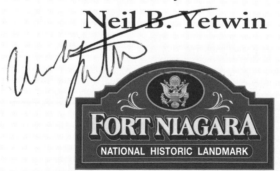

FORT NIAGARA
NATIONAL HISTORIC LANDMARK

• OLD FORT NIAGARA ASSOCIATION •
FORT NIAGARA NATIONAL HISTORIC LANDMARK
Youngstown, New York

To my son... The Life & War Remembrances of Captain Mordecai Myers, 13th United States Infantry, 1812-1815...
Including Letters and Reminiscences Pertaining to His Early Life (1780-1814), and Incidents in the War of 1812

Edited, with Biographical Introduction by Neil B. Yetwin

© Old Fort Niagara Association, Inc., 2013

Published by:
Old Fort Niagara Association, Inc.
Fort Niagara National Historic Landmark
Fort Niagara State Park • Youngstown, New York, United States of America

First Edition, August 2013 • Printed in the United States of America

ISBN: 0-941967-31-X
ISBN-13: 978-0-941967-31-0

Publisher's Cataloging in Publication Data

Myers, Mordecai, 1776-1871.
    To my son... : the Life & War remembrances of captain Mordecai Myers, 13th United States Infantry, 1812-1815 : including letters and reminiscences pertaining to his early life (1780-1814), and incidents in the War of 1812 / edited by Neil B. Yetwin.

        p. cm.

    Includes bibliographical references.

Summary: War of 1812 memoirs of Mordecai Myers, Captain, 13th U.S. Infantry, United States Army, and the only Jewish-American officer serving on the Niagara Frontier during 1812-1813. Based partly on his letters, Myers describes his wartime experience, during which he was wounded. Myers turned to politics, serving in numerous positions, including New York State legislature and mayor of Schenectady, N.Y., and was a candidate for U.S. Senate.

His career also included strong leadership in the Masons. Myers was a founding organizer and original member of the Society of the War of Eighteen Hundred and Twelve, founded in 1826. He also joined the Veteran Corps of Artillery of the State of New York, and served as Brigade Major of the Corps (a New York State Militia unit) from 1825-1835.

The original reminiscences are recognized as a valuable first-person account of early campaigns on the Niagara and St. Lawrence, especially the battles of Stoney Creek and Chrysler's Farm, as well as on military life.

        ISBN-13: 978-0-941967-31-0
        ISBN: 0-941967-31-X

1. Myers, Mordecai, 1776-1871 – Personal narratives. 2. United States – History – War of 1812 – Personal narratives. 3. United States – History – War of 1812 – Participation, Jewish. 4. United States. Army – Infantry – Officers – Personal narratives.
I. Myers, Mordecai, 1780-1871. II. Yetwin, Neil B.

E361.M93 2013
973.5'2092 dc23

Front Cover : Portrait of Mordecai Myers as a Captain of Infantry in the New York Militia, c. 1810. Original oil on wood panel, by John Wesley Jarvis. *Image courtesy, Toledo Musem of Art; 1950.275. ©Toledo Musem of Art.*
Back Cover : "Peace." Ink and watercolor drawing by John Rubens Smith, c. 1815. In an allegory of the Treaty of Ghent, signed on Dec. 24, 1814, America and Britannia hold olive branches before an altar; Sailors hold American and British flags and a white banner of mutual peace; and, above it all, a dove of Peace. *Courtesy, Library of Congress; Reproduction Number: LC-USZC4-3675. Digital restoration for publication by Harry M. DeBan, Old Fort Niagara.*

CONTENTS

To Rosy,
who has always kept me human,
all too human throughout
all of my encounters with
civilization and its discontents,
and in all of the stages
on life's way.

Acknowledgements

I would like to thank the dedicated staffs of the many libraries, archives, city halls, county clerks' offices, courthouses, and historical societies who provided access to their holdings and granted permission for their use in this book. They are cited thoughout the Notes. I am also indebted to the following individuals whose assistance made this work possible: *Dr. Allan Boudreau*, former Director of the Library of the Grand Lodge of New York; Carolyn Castor, *Dr. William R. Marsh* and the late *Alva Middleton*, collateral descendants of Mordecai Myers who graciously shared their private family papers and genealogical research into the Myers family history; *Harry M. De Ban*, Publications Director of the Old Fort Niagara Association, for his knowledge, patience and editorial skills in fashioning the concept of this book into reality; *Katherine Delain*, a tireless advocate for the improvement of Schenectady's cultural and civic life, whose interviews with me on cable network SACC-TV promoted local awareness of Mordecai Myers's historical significance; the late *Professor Emeritus Arthur A. Ekirch, Jr.*, of the State University of New York at Albany, whose personal encouragement and thought-provoking graduate courses in American Intellectual History and Classic American Historians helped to inspire this and other projects; *Ellen Fladger*, Head of Special Collections at the Schaffer Library of Union College, Schenectady, New York, guided me through the sources of Mordecai Myers' personal and political life in Schenectady; *Donald E. Graves*, one of Canada's foremost military historians and Managing Director of the Ensign Heritage Group, generously answered my many and varied questions about obscure aspects of the War of 1812 and opened the door to the possibilities of this book; *Paul Grondahl*, author and award-winning *Albany Times Union* journalist whose early coverage of my research sparked interest in Mordecai Myers and introduced his story to a wider audience; *Heather Halliday* and *Susan Wood-*

land of the American Jewish Historical Society for alerting me to the existence of little known Mordecai Myers-Benson J. Lossing documents; the late *Dr. John S. Hoes,* a 7th generation direct descendant of Mordecai Myers, for his hospitality and granting of unlimited use of his family papers, images and memorabilia; the late *Dr. Jacob Rader Marcus,* founder of the American Jewish Archives, for his personal correspondence and suggestions; the late *Walter Mason,* owner of the Man Homestead at Constable, New York, who gave me a complete tour of his home and allowed me to roam his property at will, despite my having arrived as an unannounced stranger on his doorstep; *Pamela A. Parish,* who translated several French documents relating to Colonel Irenee Amelot de LaCroix; *David S. Scannell,* for providing documents from the Albon Man Papers; the late *Rabbi Emeritus Michael Szenes* of Schenectady's Congregation Gates of Heaven, for permission to use the synagogue archives; *Wayne Somers,* proprietor of Hammer Mountain Books, Schenectady, New York, whose advice and unflagging interest over many years made the conducting of my research an even more fulfilling intellectual odyssey; the late *Malcolm Stern,* dean of American-Jewish genealogy, who directed me to several key sources; the employees of Schenectady's Vale Cemetery and the volunteers who donate their valuable time to the Vale Cemetery Association; and *Tracy Varites,* Program Manager for the New York Council for the Humanities, who has continued to provide enthusiastic support for my presentations about Mordecai Myers throughout New York State.

A special thanks to my friends and colleagues at Schenectady High School for their boundless energy, interest and humor; to my children, *Alexander, Arik* and *Hannah,* for their patience with me and for allowing Mordecai Myers to share our home for nearly all of their formative years; and especially to my wife and lifelong partner *Rosaline Eva Horowitz,* whose love, patience and critical skills remain a constant source of strength and inspiration.

Neil B. Yetwin
Schenectady, New York
Summer, 2013

Foreword

I am very pleased to be able to contribute a few words to introduce this splendid new edition of the War of 1812 memoirs of Captain Mordecai Myers of the *13th U.S. Infantry*. Myers has always been one of my favorite American eyewitnesses of the military operations in the northern theatre of war and, as his memoirs demonstrate, he was an intelligent, humane and energetic officer in wartime, and a public-spirited citizen in peacetime. Myers was also fairly unique, not only because of his religious faith — as Jewish officers were few and far between in the United States Army of the time — but also because he actually took the trouble to obtain some formal education in military matters before going on active service.

Mordecai Myers was not a young man when he went to war, and it is interesting to compare his reminiscences of the Niagara Campaign of 1813 with those of two younger officers who fought in the same campaign and whose memoirs have recently been edited for publication.* These are Lieutenant William Jenkins Worth of the U.S. Army and Lieutenant John Le Couteur of the British Army, both 19 years of age. Le Couteur and Worth do not downplay the danger and the misery of campaigning, but one gets the impression that both young men thought the whole thing a great adventure. The memories of Mordecai Myers, in contrast, are more sober, patriotic but not bombastic, and full of interesting detail which makes them an excellent historical resource.

If Neil Yetwin had simply annotated his subject's wartime reminiscences and left it at that, he would have accomplished a great deal, but he has done much more. Through impressive research Yetwin has set Mordecai Myers in his proper historical and personal context by reconstructing the history of the Myers family and, in doing so, provided much information about the life of the Jewish minority group in the time of the early republic. The Myers family is not only interesting because they were Loyalists who left the new United States at the end of the Revolutionary War, but also because they were part of a not inconsiderable group of Loyalists who returned to live in the republic. Yetwin has also done a magnificent job of annotating his subject's material. Frankly, as someone who has edited a number of historical manuscripts for publication, I cannot think of any further scholarship that might have been done to bring this work to the level of publication.

The result is that the wartime memoirs of Mordecai Myers — as well as additional materials relating to the man — have been placed in proper context and thoroughly annotated. Neil Yetwin is to be complimented for his impressive industry, and the Old Fort Niagara Association's Publications Program for making this important historical resource available to a new audience.

> Donald E. Graves
> "Maple Cottage"
> Valley of the Mississippi, Upper Canada
> Victoria Day 2013

————————

*Donald E. Graves, editor, *Merry Hearts Make Light Days: The War of 1812 Journal of Lieutenant John Le Couteur, 104th Foot* (Montreal, 2012) and *First Campaign of an A.D.C.: The War of 1812 Memoir of Lieutenant William Jenkins Worth, United States Army* (Youngstown, New York, 2012).

Publisher's Note:
The annotations for the "Brief History" Introduction which follows, are placed as Foot Notes.
However, due to the frequency and length of many of the annotations for "Part One" and "Part Two", they appear as End Notes at the close of their respective main text.

An Act Declaring War Between the United Kingdom of Great Britain and Ireland and the Dependencies thereof, and the United States of America and their territories.

Be it enacted by the Senate and House of Representatives of the United States of America in Congress assembled, That war be and the same is hereby declared to exist between the United Kingdom of Great Britain and Ireland and the dependencies thereof, and the United States of America and their territories; and that the President of the United States is hereby authorized to use the whole land and naval force of the United States to carry the same into effect, and to issue to private armed vessels of the United States commissions or letters of marque and general reprisal, in such form as he shall think proper, and under the seal of the United States, against the vessels, goods, and effects of the government of the said United Kingdom of Great Britain and Ireland, and the subjects thereof.

APPROVED, June 18, 1812.

Anatomy of a Declaration of War

• June 1, 1812: President James Madison send a message to Congress regarding ongoing hostile actions by Great Britain against the United States.

• June 4, 1812: The United States House of Representatives approves a Declaration of War by a vote of 79 yeas to 49 nays.

• June 18, 1812: The United States Senate votes for war by a vote of 19 yeas to 13 nays. President Madison signs the Congressional declaration of war, marking the beginning of the War of 1812.

• June 19th, 1812: The *Declaration of War* "Proclaimed" by President Madison and certified by Secretary of State James Monroe, completing the formal legalities of the process under United States law.

(*United States Statutes at Large,* Twelfth Congress, Session One, Chapter 102, Page 755; Statute 1, June 18, 1812)

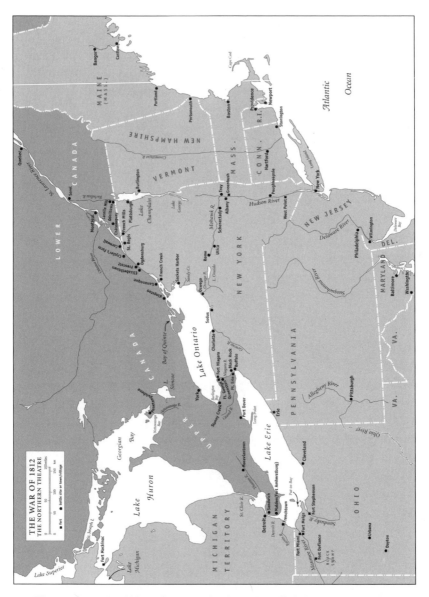

Throughout the War of 1812, the heaviest fighting occurred in
the northern border region of the United States, from Lake
Champlain to Lake Michigan.
Map from Donald E. Graves,
Field of Glory: The Battle of Crysler's Farm, 1813 (Toronto, 1999).
Courtesy, Donald E. Graves.

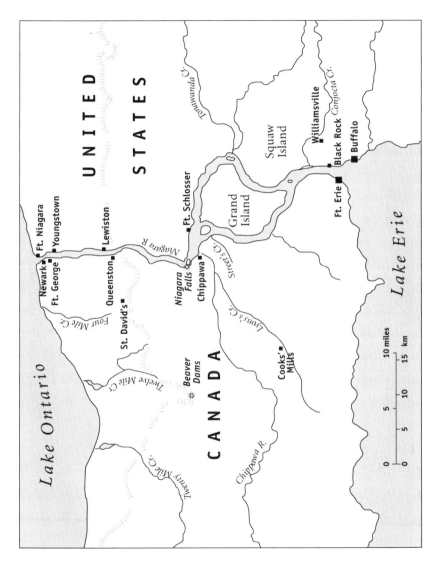

The Niagara Peninsula. During the War of 1812, both sides of
the Niagara River witnessed some of the conflict's heaviest fighting.
Map from Donald E. Graves, *Where Right and Glory Lead: The
Battle of Lundy's Lane, 1814* (Toronto, 1997).
Courtesy, Donald E. Graves

"M. M.
*Severely wounded
at Chrysler's Field
in Canada,
when Capt. of the
13th U.S. Infantry,
in the war of 1812.*

*Represented New York City
for many years in the
State Legislature.
Mayor of this City*
[Schenectady]
*& Grand Master of Masons
of the State of New York.*

*The Historic Pen hath writ his
spotless name Among the
good and brave,
the men of
deathless
fame.*"

Myers's Reminiscences: A Brief History

by

Neil B. Yetwin

By the time Mordecai Myers began writing the letters that would form the basis of his posthumously-published *Reminiscences,* he had already lived several lives and enjoyed a well-earned reputation throughout New York State as a military hero, principled merchant and landlord, tireless assemblyman and devoted Freemason. Along the way he had won and lost two fortunes and fathered ten children, one of whom became the great-grandfather of Pulitzer Prize-winning poet Robert Lowell (1917-1977).[1]

Myers's memoir was privately printed in 1900 by the Crane Company of Washington, D.C. in a limited run of 100 softbound copies under the title, *Reminiscences 1780 to 1814 Including Incidents In The War of 1812-14; Letters Pertaining to His Early Life, Written by MAJOR MYERS, 13th Infantry, U.S. Army To His Son.* Like many of the self-made men of his time, Myers had little if any formal education, so the idiosyncratic spelling, capitalization and punctuation common in his surviving handwritten correspondence were likely corrected by his eldest surviving son Theodorus. But when Theodorus died in 1888 the task of publishing the memoir fell to his daughter, Cassie Mason Myers Julian-James (1851-1922). Mrs. Julian James was a Washington, D.C. philanthropist and vice-president of Chapter 3 of the Colonial Dames of America who helped oversee the extensive costume collection at the Smithsonian Institute's National Museum of American History.[2]

---

(1) Myers's son, Algernon Sydney Myers (1829-1883), had one daughter, Kate, who married Union College professor Robert Traill Spence Lowell, the poet's grandfather. For Robert Lowell's unique interpretation of his great-grandfather, see his essay, "91 Revere Street" in Robert Lowell, *Life Studies* (New York: Vintage Books, 1959), pp. 11-13.

(2) James, Catalina Mason Myers Julian. *Biographical Sketches of the Bailey-Myers-Mason Families, 1776-1905* (Washington, D.C., 1908), pp. 37-58.

The *Reminiscences* consist of the author's autobiographical recollections from about 1780, when he was four years of age, to
about 1840. The frontispiece is a portrait of Myers dressed in the
uniform of a *New York State Militia* Captain, completed in 1810
by John Wesley Jarvis.[3] Beneath the portrait is a facsimile of
Myers's distinctive signature. In addition to the main body of
the 56-page work, there are additional extracts of letters in which
Myers revisited his war years and are presumably taken from a
letter or letters written after February 1853. There are also copies of two pieces of correspondence describing an affair of honor
in which Myers acted as second. The whole is followed by a brief
postscript written either by Cassie or Theodorus, noting Myers's
tenure as Mayor of Schenectady, New York, and his having served
as Grand Master of the Grand lodge of the State of New York.
The final piece is an abbreviated version of Myers's lengthy obituary.[4] The factual errors that appear in the *Reminiscences,* such as
dates, locations, or names of those with whom he served were
probably due to Myers's age (76) at the time of the writing, or the
simple passage of time which often causes memory to dim. At
least one such error was corrected prior to publication.[5] Eleven
of the original one hundred copies, which were distributed to

---

(3) Jarvis' work was in great demand in 1810 when he maintained a studio on Wall Street in
Manhattan. He created many now-familiar portraits of well-known political, literary and
military figures of his time, such as John C. Calhoun, De Witt Clinton, James Fenimore
Cooper and Oliver Hazard Perry; his portrait of Andrew Jackson, which adorns the American
$20 bill, was part of a commissioned series he did on heroes of the War of 1812. Jarvis also
served as an ensign, then lieutenant under Captain Mordecai Myers in the First Brigade, *Third
Regiment of New York Infantry* during 1810-1811, at which time the men became acquainted.
The Jarvis portrait of Myers is on display at the Toledo Museum of Art. There exist two copies;
one is on display at the Vanderpoel House in Kinderhook, New York (as part of the Columbia
County Historical Society), the other remains in private hands. Harold E. Dickson, *John
Wesley Jarvis: American Painter, 1780-1840. With a Checklist of His Works* (New York Historical
Society, 1949), p. 86, pp. 96-106, pp. 341-83; Hastings, Hugh, ed., *Military Minutes of the
Council of Appointment of the State of New York, 1783-1821* (Albany: James B. Lyon, 1901. Vol.
II) pp. 1157-58, p. 1175, p. 1299; *Museum Notes,* Toledo Museum of Art, 1957, pp. xliv-xlv.

(4) The obituary appeared in the *Schenectady Evening Star* of January 23, 1871, p.3.

(5) Myers wrote "Colonel Butler" on pp. 36-37, which was corrected to read "Boerstler" (Lt.
Col, later Col. Charles Boerstler) under "Errata" on an un-numbered page at the end of the
Reminscences.

family members, friends and several repositories, are currently known to have survived. Ten are housed in the collections of the libraries at Harvard, Columbia and Indiana Universities, Union College, the University of Rochester, the U.S. Army Military History Institute, The Navy Department, the Wisconsin Historical Society, the New-York Historical Society, the Schenectady County Historical Society, and the Boston Public Library. One copy remains in private hands.

In recent years several historians of the War of 1812 have cited Myers's narrative as an important first-person account from the American side of the Niagara and St. Lawrence Campaigns, and recognized its author's professional qualifications and attention to duty. John C. Fredriksen has described Myers' s work as "an excellent account of army life." [6] James and Nicko Elliott point out in their retelling of the Battle of Stoney Creek that "with several years of militia service and training at a private military academy before the war, Myers was unusually qualified for his regular army commission." [7] And in his definitive work on the Battle of Crysler's Farm, Donald E. Graves characterizes Myers as a "representative of the better type of company officer in the northern army" who "served with distinction" as one of a group of "excellent unit officers." [8]

Myers's *Reminiscences* is also unique because its author was one of a handful of Jews who served on the American side of the conflict and it has been suggested with some accuracy that he was the only Jewish officer serving on the Niagara Frontier during the winter of 1812-1813.[9] Yet several historians of the

---

(6) Fredericksen, John C. *The War of 1812 in Person: Fifteen Accounts by U.S. Regulars, Volunteers and Militiamen* (Jefferson, N.C.: McFarland & Company, Inc. Publishers, 2010), p. 29, #86.

(7) Elliott, *Strange Fatality*, p. 150.

(8) Graves, Donald E. *Field of Glory: The Battle of Crysler's Farm, 1813* (Toronto, ON: Robin Brass Studio, 1999), pp. 12, 39, 320.

(9) Leon Huhner, "Jews in the War of 1812", *American Jewish Historical Society*, XXVII (1918), pp. 173-200. Hereafter cited as Huhner; Selig Adler, *From Ararat to Suburbia: The History of the Jewish Community of Buffalo*. (Philadelphia: The Jewish Publication Society of America, 1960), p. 402. Hereafter cited as Adler.

American-Jewish experience have found the work wanting, claiming that Myers's descendants, who were not raised as Jews, edited the memoir in an attempt to purposely obscure the author's — and thus part of their own — ancestry.[10] It is true that the *Reminiscences* contain no references to Myers's Jewish background, but neither does his surviving correspondence. What these omissions may indicate is that like many American Jews of his generation, Myers saw himself as a patriotic American first and foremost and wished to be perceived as one. He never denied his Jewish background or converted to Christianity, but instead subscribed to the democratic ideals of the new republic, to the Constitution that guaranteed his basic freedoms, and to the spirit of universal toleration that is one of the foundations of American Freemasonry. Myers often employed Masonic language to express his religious beliefs, once describing the one God to whom he prayed as "the Great Architect, the universal Parent and Protector of all Mankind."[11] The original correspondence that formed the basis for the *Reminiscences* has not been located, so accurately verifying any claims to its completeness is problematic. Yet its content, language, structure and overall tone is completely consistent with the surviving correspondence and drafts of public statements that Myers left in his own hand. The *Reminiscences,* then, is in all likelihood an accurate, uncensored transcription of the original letters, and not the work of a descendant intent on recasting its author's ethnic background.

Myers maintained lifelong social, political, commercial and Masonic associations that together provided an essential network in his rise from poverty and obscurity. None, however, was as important to him as his links to the American military. From his early years in the Virginia and New York State Militias to his captaincy in the regular army during the war, Myers clearly enjoyed commanding his company of the *13th Infantry Regiment,*

(10) Huhner, 177; Jacob Rader Marcus, *Memoirs of American Jews 1775-1865.* (Philadelphia: The Jewish Publication Society of America, 1955. Volume I), p. 50; Marcus, *United States Jewry,* Volume 1, p. 97; Adler, p. 5.

(11) *Biographical Sketches,* pp. 15-16.

and relished the camaraderie, adventure and pageantry of military life. It may be telling that in a memoir of just 56 pages, 33 deal exclusively with his military experiences. He did harbor some bitterness over his discharge and failure to secure what he believed would have been a well-deserved promotion, but this never compromised his sense of patriotism or his respect for the military establishment.[12] In fact, he often wistfully described his post-war civilian endeavors — including those that met with success — as being among "the Comman [sic] walks of Life" in comparison with life in the army.[13]

Myers recognized that the most decisive moment of his life occurred on the afternoon of November 11, 1813, when he sustained severe injuries at the Battle of Crysler's Farm. Had he not been wounded that day and not been discharged at the war's end, or been presented with the opportunity for promotion, it is reasonable to assume he would have remained in the army and pursued a military career. Nevertheless, he sensed quite clearly that it was the events of that day that allowed him entrance into a prominent, well-connected New York family and helped steer him toward what would be a noted political career. So significant was Crysler's Farm to Myers that he purchased a silver soup ladle and had engraved upon it his initials and the date "November 11, 1813", in much the same way it would be immortalized on the obelisk marking the Myers family plot in Schenectady's Vale Cemetery.[14] To the end of his days, "the Major," as he was known, would continue to take his greatest pride in having fought bravely during the War of 1812 as Captain Mordecai Myers of the *13th United States Infantry*.

---

(12) See editor's introduction for Myers's post-war participation in the Military Society of the War of 1812, his efforts on behalf of veterans while serving in the New York State Assembly, and his role in supporting the New York State Militia.

(13) Myers to Winfield Scott and Isaac Varian, November 16, 1841. Mordecai Myers. *Letterbook, 1835-1841.* Courtesy of the late Dr. John S. Hoes.

(14) The ladle remains in the possession of Myers's descendants.

Mordecai Myers, U. S. Army: War of 1812

Portrait of Mordecai Myers as a Captain of Infantry in the New York Militia, was painted c. 1810. Original is an oil on wood panel, by John Wesley Jarvis. The original is 86.1 cm high, by 67.7 cm wide, and is owned by the the Toledo Museum of Art, Toledo, Ohio. It was purchased in 1950 with funds from the Florence Scott Libbey Bequest, in memory of her Father, Maurice A. Scott. Jarvis was born in England, but came to the United States with his family as a child. *Courtesy, Toledo Musem of Art; 1950.275.*

Part One:
The Life & Times of
Mordecai Myers

by
Neil B. Yetwin

1. *"From the Rhineland to America, 1095 – 1757"*

On January 24, 1871, hundreds of spectators gathered near the brick house at 71 Union Street in Schenectady, New York to pay their final respects to their former mayor, Mordecai Myers. Four inches of snow had fallen with the sub-freezing temperatures the night before, but the harsh Mohawk Valley weather failed to prevent one hundred representatives from New York State's Masonic Lodges from turning out in full ranks to salute their departed Brother. The local newspapers characterized "the Major", as the deceased was known, as "a well-bred gentleman of the old school" whose "high-toned princely integrity and honor were in marked contrast to the easy virtue of these modern days." After a brief viewing and the offering of condolences to the Major's family, the Masons transported the silver-handled rosewood coffin to the Lodge Room in their Church Street Temple, where they conducted a service in accordance with their ancient Craft. The funeral procession then reformed, crossed the State Street Bridge over the Erie Canal, and moved two miles up State Street to the entrance of Vale Cemetery. There, later that day, it was interred in its final resting place in the Myers family plot.[1]

Today the twelve-foot-high granite monument marking the Myers Family plot stands in an overgrown, mossy glade deep in one of the Vale's oldest sections, struck only occasionally by a few persistent shafts of sunlight. The monument dutifully lists the names and appropriate dates for the Major and his immediate family, as well as his well-known public achievements:

> *Represented New York City for many years in the State*
> *Legislature*
>
> *Mayor of this City*
>
> *Grand Master of Masons of the State of New York*

and the event that Myers himself considered the most decisive of a life that lasted for nearly a century:

> *Severely wounded at Chrysler's Field in Canada, when*
> *Capt. of the 13th U.S. Infantry, in the war of 1812*

Yet, most of Schenectady's citizens, as well as many members of his own family, remained unaware of many of the details of Myers's early life.

Eighteen years prior to his death, Mordecai Myers began his *Reminiscences* by informing us that his father "was a Hungarian" and his mother "an Austrian by birth."[2] He failed to mention his parents' first names, nor did he specify their Jewish ancestry. But his identification of their national origins along with the surname "Myers" suggests significant clues to his family's history.

The Jews who survived the Roman destruction of Jerusalem in 70 C.E. were dispersed along the Mediterranean trade routes into Europe, North Africa, and the Iberian and Italian Peninsulas. Those who settled in Germany called their new home "Ashkenaz". Over time, the surname "Myers", originally the first name of a remote ancestor, became a common Ashkenazi name throughout the Jewish enclaves of Southern and Western German.[3] Crusader armies entering the Rhineland in 1096 drove thousands of the Ashkenazi Jews into Hungary; a second German-Jewish immigration into Hungary occurred after they were invited to settle in those areas severely depopulated by the Mongol invasion of 1241. Hungary's rulers hoped that their presence would help revive the economy.[4]

Hungary was then subdivided into "Three Hungaries" after Sulieman's conquest of the region in 1526, one of which included the Hapsburg Dynasty's Royal Territories. All of Hungary's Jewish communities became permanent possessions of the Hapsburg Empire when Turkey relinquished its Hungarian holdings to the Hapsburg Emperor Leopold I in 1699. The Hungarian Jews were treated less harshly than the Jews in other European countries, but Leopold preferred his dominions to be Catholic and persecuted the Jews along with Protestants and dissenters. He only allowed the Jews to live relatively undisturbed because the exorbitant rents, taxes and levies imposed upon them were a reliable source of income for his empire.[5] By the early 18th century, substantial numbers of Hungarian Jews, nearly half of whom were of Austro-German descent, had settled in Western Hungary's so-called "Seven Communities" along the Austrian border.

On October 26, 1726, Leopold's successor Charles VI issued the "Familientz" or "Familiants" Laws, which stated that only the eldest male member of each Jewish family could marry, have a family and be considered a legal resident of the country. All others would be treated as foreigners who could only marry abroad, and families that only produced daughters would become extinct, which in turn forced large numbers of young Jews wishing to marry to cross the Hungarian-Austrian border.[6] The prospects for the Hungarian Jews appeared no brighter when Charles' daughter Maria Theresa inherited the Hapsburg Throne. She continued the persecution of dissenters and non-Catholics, but reserved a particular contempt for the Jews: "I know of no worse public plague than this people," she once wrote.[7]

Benjamin Myers, Mordecai Myers's father, may have been prevented from marrying under the Familiants Laws if he had older brothers. As a Jew he would never be allowed to own land or join a guild; he would be taxed by everyone from the Emperor or Empress and nobles to the clergy and local officials; and there would always be the every-present threat of expulsion, imprisonment or, in some rare cases in Hungary, death. America promised a better life. Myers engaged a marriage broker to find a suitable wife to accompany him on his journey, and the broker soon found a young Austrian girl named Rachel, whose maiden name remains unknown. At the time of their marriage, probably sometime in 1757, Benjamin was 24 and Rachel 12 years of age.[8] While this may appear inappropriate to modern sensibilities, in mid-18th century Hungarian-Jewish communities, males 13 and over were considered adults, as were females 12 and over with parental consent. If Rachel's parents only had daughters, their family faced extinction under the Familiants Laws. Under these circumstances, Benjamin Myers must have seemed a likely prospect.[9]

The couple left the Hapsburg Dominions and made their way first to Holland, then England, and on to America.[10] They arrived several weeks later in New York, where a Jewish community had been firmly established since 1654. The new arrivals were likely given a warm and sympathetic reception, but the Myers, coming as they did from small border towns, must have been ill-prepared for Manhattan's diverse and sizeable population of 18,000 and its bustling commercial and social life. The couple may have been advised by the community's leaders to relocate to a smaller community more congenial to their immediate needs, and suggested Philipse Manor, about 20 miles up the Hudson River. Philipse Manor, today part of Yonkers, New York, was then a sparsely populated rural area of about 300 farmers, mainly of Dutch ancestry and some settlers from New England.[11]

It is possible that the Myers' New York advisors had practical reasons for suggesting Philipse Manor. The Hudson River and its tributaries enabled local settlers to visit New York City for trade and supplies. By investing in a small dry goods concern or trading post at Philipse Manor, operated by a grateful co-religionist, the city merchants might tap more efficiently into a rural market. And if Benjamin Myers engaged in some peddling among local farmers, with goods provided by these same merchants, all parties would benefit and the young immigrant-entrepreneur would learn English that much more quickly. It is not known how long the Myers' resided at Philipse Manor, but on Tuesday, December 13, 1758, the couple's first child, Benjamin, Jr. — Mordecai Myers's eldest brother — was born at Philipse Manor.[12]

Now that they had begun a family, the Myers must have come to miss the secure familiarity of a Jewish community, with ready access to a synagogue and properly prepared foods. At the same time, they wanted to avoid returning to New York City. Following a possible suggestion by their New York benefactors, they moved to Newport, Rhode Island. Newport's population was about half of New York City's, but it had a Jewish community that enjoyed close commercial and familial ties with its New York counterparts.

2. "Newport and the Revolution, 1761-1783"

Newport, Rhode Island was one of the five principal port cities (New York, Philadelphia, Savannah and Charleston were the others) of the American colonies that had small but thriving Jewish communities. When the Myers arrived there in 1761 there were 10 Jewish families, about 56 individuals who were enjoying unprecedented religious freedom as well as full participation in the town's civic, commercial and cultural life. Aaron Lopez, the wealthy "American-Jewish merchant Prince" lived there and was playing a major role in ushering in Newport's "Golden Age". [13]

Benjamin Myers must have arrived with favorable letters of introduction from New York and impressed Newport's Jewish communal leaders as a young man of worth. Two of them – Moses Levy and Jacob Isaacks – posted bond of 10,000 pounds "for the good behavior of Benjamin Myers and his family" on August 3, 1761. [14] The Myers moved into a house on Marlborough Street which would serve as both family residence and small business for the next twenty years. [15] Myers also appears to have had some learning. A Jewish traveler named "Tobiah the Levite" boarded at the Myers' home while on a fundraising mission that same year. After leaving Newport, he wrote a letter of thanks to Newport's Jewish community for their generosity, adding a special appreciation to his host, "Rabbi Meir." The term "rabbi" at that time meant a respected scholar or teacher, not necessarily the spiritual leader of a congregation. [16] And when Newport's Jewish commercial elite formed what has been called "America's first Jewish club" on November 25, 1761, Benjamin Myers was hired as the club's steward. [17]

During the next three years, rising tensions between Great Britain and the American colonies were beginning to have a direct impact on even the largest of Newport's mercantile houses , but most especially on smaller merchants like Benjamin Myers. The Currency Act of September 1, 1764, an attempt to prevent the colonies from paying debts in England with depreciated currency, prohibited the issue of legal tender currency in all the colonies and created a money shortage. As a result, Benjamin Myers was forced to declare bankruptcy on March 26, 1764.[18] The town's Jewish community came forward to offer the family help. From September 3, 1764, through November 25, 1776, the Myers' meager income was regularly supplemented with such essentials as food, firewood, whale oil and hay, as well as goods the family could sell and occasional gifts of cash. Some of these were charged to the community's charitable fund, while others were charged personally to Aaron Lopez. Benjamin Myers was also hired as the community's "shochet" or ritual slaughterer responsible for preparing meat according to Jewish tradition.[19]

By 1769 Newport's Jewish population had grown to 25 families (about 121 individuals) in a total population of just over 9200.[20] In order to accommodate the needs of this growing community, Benjamin was hired as the sexton of Newport's synagogue, Yeshuat Israel. As such he was required to attend all weddings, funerals and circumcisions, keep the synagogue clean, its locks in working order, and its traditional candles appropriately lit. Myers may also have taught prayers, ceremonies and bar-mitzvah preparation to children in the small annex attached to the synagogue's main sanctuary. [21]

Benjamin and Rachel Myers had eight children by 1774,[22] so it is small wonder that Savannah merchant Philip Moses, who had visited Newport on business the year before, sent funds on February 1, 1773, for the sole purpose of "assisting Mr. Myers Familys (sic)." [23] The Myers began to take in more boarders that year for extra income. One was a Portuguese-born tailor from New York named Isaac Nunez Cardozo, great-grandfather of future United States Supreme Court Justice Benjamin Cardozo.[24]

The impending threat of war between Great Britain and her colonies continued to make living conditions in Newport more unstable. An October 1775 threat to bombard the town, announced by the commander of the British fleet in Newport Harbor, caused half the town's population to evacuate. To make matters worse for the family, Benjamin Myers had suffered an accident the previous March which resulted in a broken arm that failed to heal correctly and caused his health to slowly decline. But on May 31, 1776, Rachel Myers gave birth to her ninth and last child whom they named Mordecai. Benjamin Myers continued to fade and finally died on November 20, 1776, about six months after Mordecai was born.[25]

Shortly after Mordecai Myers's birth, 800 American troops stationed on Rhode Island marched into Newport and held the town while the British fleet occupied the harbor. Anyone suspected of loyalty to the Crown was forced to sign a loyalty oath to the American cause or face harsh reprisals. Benjamin Myers, Jr. and his younger brother Abraham chose to remain loyal to the Crown and fled Newport, eventually joining Van Cortlandt Skinner's New Jersey Volunteers, or "Skinner's Greens" for the color of their uniforms.[26]

There were about 2500 Jews in the American colonies at that time, about one-tenth of 1% of the population. The vast majority of these supported the American cause, though a handful of the more affluent families remained loyal to England. Rachel Myers, fearful for her future and probably urged on by her older sons, became one of the few poorer Jews who became Loyalists. When a force of 8000 British troops occupied Newport on December 6-8, 1776, all of its pro-independence citizens were allowed to leave and the town was soon repopulated by Loyalists and British soldiers. But factories closed, commerce came to a standstill, and nearly all the Jews evacuated, leaving Rachel and her seven remaining younger children with few resources. Food and fuel shortages became so acute that after 10,000 American troops and Count D'Estaing's French fleet forced the British out of Newport, Rachel was forced to take her family to New York City, "the capital of Loyalist America" since its occupation by the British on September 15, 1776. [27]

3. *"I am now Reduced to the greatest extremity of poverty and Pinching Want': The Exile to Nova Scotia and New Brunswick, 1783-1787."*

The Myers family move to New York City was part of a larger ingathering of thousands of Loyalists from throughout the colonies that soon overwhelmed the British authorities. General William Howe, who was in charge of the city, did what he could for the Loyalists by way of provisions, cash and even lotteries, but escalating unemployment and shortages of food and housing stretched the available resources to their limits. In desperation, Rachel Myers petitioned for assistance on April 3, 1781, citing her sons' loyalty, her forced exile from Newport, and her attempts to support her family. "But all her industry is not now sufficient to afford them the necessaries of life," she stated, "which constrains her to implore your Excellency to extend her some relief from government, by permitting her to receive for her family such rations of provisions, etc. as may be thought necessary." The family finally began receiving provisions on June 12 and they remained in the city for the duration of the war. [28]

As the war drew to a close, the British government announced that land grants in Nova Scotia were to be offered to the Loyalists as compensation for their losses, services and suffering. Rachel, Benjamin and Abraham, fearing post-war anti-Loyalist reprisals, decided to move the family to Nova Scotia. Throughout the spring of 1783, twenty British vessels collected thousands of Loyalists from New York City, Long Island and Staten Island and assembled them at Sandy Hook, New Jersey to await transport to their new refuge in Canada. In time, four separate fleets would carry approximately 50,000 Loyalists to Nova Scotia. The Myers left with the "First" or "Spring Fleet" in May, 1783.[29]

The Spring Fleet arrived at Parrtown (later Saint John, New Brunswick) at the mouth of the Saint John River on May 18th, since celebrated annually as "Landing Day." John Parr, who had been appointed governor of Nova Scotia, had delayed the regaining of old land titles that would have made for a smoother distribution of lots for Loyalist families. To make matters worse, the spring of 1783 was cold and wet; there were food and water short-ages, swarms of insects, poor sanitation and rising alcoholism, theft and violence among the new arrivals. Additional fleets flooded Parrtown with even more refugees. Parr had to admit by January 1784 that the refugees were in a "wretched condition" and "destitute of almost everything." Seven-year-old Mordecai and his family, along with thousands of others, languished that winter in a makeshift city of tents and crude huts.[30] Some relief came in early summer of 1784 when Benjamin, Abraham and 65 other veterans of the Loyalist Provincial Corps were each granted 200 acre lots up the St. John River surrounding Coy Lake. [31]

Governor Guy Carleton arrived at Parrtown in November 1784 to replace the unpopular Parr and saw to it that 2.5 million acres of old Nova Scotia land titles were escheated (voided) in order to expedite land distribution to the Loyalists. Unfortunately, much of the best land had already been granted to those with political influence or funds for surveys. Benjamin and Abraham went upriver to Coy Lake to begin developing their tracts before another long winter arrived, while the rest of the family waited at Parrtown. During that period Rachel Myers initiated the first in a series of petitions for land and aid that clearly reflect her despair and that of most of her fellow "distressed Loyalists." She poignantly described her "Distressed situation", "fatherless Children" and "Real Needy Family" when asking Governor Carleton to grant her a farm. [32] Yet four months later she was "now Reduced to the greatest extremity of poverty and pinching Want" and implored the governor "to Notice my distressed Situation as being a widow and a Large family of Fatherless Children to provide for...." She needed the farm, she insisted, "to prevent myself and family from suffering with hunger and Nakedness...." [33]

Rachel was finally granted a lot at Grand Lake, but it was so poor that she had to apply for another 15 acres on Grimross Neck. That tract proved "to (sic) Low for Bilding (sic) on", as did the family's next lot at Hampton Springs.[34] Benjamin interceded on his mother's behalf, drawing a half-acre lot in the Town Plot at Parrtown.[35] One month later, Rachel exchanged that lot for one belonging to Samuel Dickinson.[36] These types of informal, undocumented property transactions among the Saint John River Valley Loyalists also contributed to their resettlement difficulties.

By 1786 it was becoming increasingly clear to the Myers family that their lands would yield little if anything to sustain them. The severe winter of 1786-87 had been particularly hard on the Saint John River settlers, more so for urban families like the Myers who lacked the necessary skills for survival in a primeval wilderness. It was also the family's third year in what was now New Brunswick, which had been created from Nova Scotia in 1784 with the goal of alleviating some of the Loyalists' hardships. British policy provided the refugees with full provisions for the first year and two-thirds for the second year of settlement, which meant that most of the families were now receiving only one-third of their original allotment.[37]

At the same time, American was experiencing a period of post-war reconciliation. Many of the anti-Loyalist measures that were in place before and during the Revolution were barely enforced, and by 1787-88 were repealed altogether. Correspondence between family members and ethnic and fraternal organizations increased, and old political differences were gradually being set aside. Returning Loyalists were generally treated cordially and the "re-Exodus" of 1787-91 witnessed the return of about 19,400 New Brunswickers to America, causing the abandonment of many of the St.John settlements.[38] On May 1, 1787, almost exactly four years after her family's arrival at Parrtown, Rachel Myers conveyed her Town Lot to Samuel Dickenson and took Mordecai, now 11 years old, and his siblings back to New York City.[39]

4. "The Militia, the Lodge, and the Tammany Society, 1794-1811"

The Myers returned to New York City in the summer of 1787 and greeted by the members of Congregation Shearith Israel, whose leaders had first received Rachel and her late husband Benjamin to America thirty years earlier. Anxious to show their gratitude for allowing them to put their Loyalist past behind them and start anew, the family made contributions to the synagogue's charitable fund as well as a fund for the construction of a new mortuary chapel.[40] Rachel even ran (unsuccessfully) for the position of synagogue "shammash", a combination secretary, messenger, sexton and charity collector.[41] Perhaps it was Rachel who instilled in Mordecai an interest in politics.

Benjamin and Abraham Myers were struggling to establish themselves as merchants and brokers in 1791 when New York was plunged into a brief but ruinous depression. Thousands of New Yorkers lost their businesses, farms and life savings through reckless speculation in unstable new banks and bank stocks, so the two brothers temporarily relocated to Richmond, Virginia, taking 15-year-old Mordecai with them.[42]

Land speculation and growing trade with the farmers of the western valleys were drawing settlers into Richmond in the late 18th century, so that by 1790 its population was about 2000. Among them was a Jewish community of twenty families, whose heads were merchants, storeowners and tavern operators. Like the Jews of Newport, they were well assimilated into Richmond's social, cultural and civic life, and kept in contact with the other Jewish communities along the eastern seaboard. Several of these men were active in the local militia and Masonic lodges.[43]

When Richmond's Congregation Beth Shalome was organized in 1791-92, the three Myers brothers were among its 29 founding members. [44] Mordecai, not quite the required minimum age of 16 for militia service but tall for his age, was allowed to serve in the 19th Richmond Regiment under Lieutenant Colonel (later U.S. Supreme Court Chief Justice John Marshall), who then commanded the Henrico County Militia. [45] The Myers brothers were back in New York before the end of 1792, having done well enough in Richmond to purchase three private seats at Shearith Israel.[46] Mordecai must have noticed while in Richmond that association with a Masonic lodge was an essential step for any ambitious young merchant. The American lodges, unlike those in Europe, welcomed Jews into their ranks, and it was the institution where many of them learned parliamentary procedure, the social graces, and connected them to a wider network that helped them enter the growing middle class. [47] As soon as he was of age in 1795, Mordecai was initiated into New York City's Phoenix Lodge No. 11. [48]

The next few years were busy ones for Myers. In addition to joining a Masonic lodge, Mordecai became a broker,[49] joined the Society of Tammany,[50] which had become Aaron Burr's Democratic-Republican political organization and, in partnership with older brother Abraham, became a state-licensed auctioneer.[51] Myers also served the first of what would be six terms as a trustee of Congregation Shearith Israel.[52] Less than three years later he was admitted into the Manumission Society of New York City, which promoted the abolition of the slave trade, the protection of free blacks from enslavement, and provided aid to those still enslaved.[53] He also enlisted in militia Captain John Swartout's Artillery Company and remained for six years until his appointment as a Lieutenant in Major James Cheatham's 3rd Infantry Regiment of the First Brigade.[54] Myers was then promoted to the rank of Senior Captain and commanded a battalion of Lt. Col. Beekman Van Beuren's Third Regiment of the New York State Militia.[55] "I became senior Captain of the regiment, and had command of a battalion," he recalled. "But before the commissions were made out for promotions (Major in my case) I left the regiment." Daniel D. Tompkins, influential in Democratic political circles and a personal friend of Myers, suggested that he study military tactics under French emigre Colonel Irenee Amelot de LaCroix, which Myers did at his own expense.[56]

Now confident and more than a little ambitious, Myers applied to Secretary of State James Madison on June 3, 1808, for what he called a "Consulship either in India on the coast of Barbary, or in the West Indies", pointing out that he was "well acquainted with all the Public Characters in this State...." [57] It was not unusual for well-connected Jewish merchants who conducted business in foreign ports to become consular agents in those ports, but Myers was a local businessman with few substantial contacts outside the city, so he failed to receive the appointment. This setback did not, however, prevent his continued rise into the highest ranks of Freemasonry. In an elaborate ceremony conducted on November 24, 1808, Myers and four others, including Daniel D. Tompkins (soon-to-be Governor of New York State and later U.S. Vice-President) became 32nd Degree Freemasons.[58]

Myers continued working as a broker and auctioneer, sometimes on his own and sometimes in partnership with co-religionists Aron Judah or Naphtali Phillips.[59] Like many of his fellow merchants, brokers and auctioneers, he learned the intricacies of the law by suing, being sued, or serving on juries, all common occurrences in the newly evolving economy. Myers's suits tended to center on unpaid debts, bills, loans, or back rent,[60] but one of the thirty cases in which Myers is mentioned reveals that he had a volatile temperament. On February 10, 1809, he was arrested for a physical attack on New York merchant Thomas Aspin. According to the court records, Myers, "with force and arms, to wit, with swords, clubs, hands and sticks, made an assault upon the said Thomas Aspin and did then and there beat, wound and ill treat, so that his life was greatly endangered." What precipitated Myers's attack is not known, but in an age when men settled their personal or political differences through duels or street brawls, it was not an unusual occurrence. There is no record of a resolution in the case.[61]

The brokerage and auctioneering house of "Myers & Judah", located at 158 Pearl Street[62] had accumulated considerable debts over the years. Their business, like hundreds of others, had been devastated by the 1807 Embargo Act, which forbade American vessels to sail for foreign ports. Thirteen hundred men had been imprisoned for debt in New York City alone. Within three years the partners found themselves in debt to 19 creditors for the enormous sum of $20,770 — not including another $3,000 owed to the Mechanics Bank — and were forced to declare bankruptcy. Their insolvency hearing was held in Albany before New York State Chancellor James Kent. By the time the hearing concluded on April 19, 1811, Mordecai Myers — broker, auctioneer, synagogue trustee, New York State Militia Captain, 32nd Degree Freemason and longtime member of the Society of Tammany — had only $225 worth of personal property to his name.[63]

Plagued by lawsuits, bankrupt and facing a bleak financial future, Myers rejoined the Third Regiment of the First Infantry Brigade (New York State Militia) as a Captain, and planned his transition from the State Militia to the regular army, just as war between Great Britain and the United States was about to be declared.[64] The New York State Council of Appointment recorded his departure from the State Militia on May 23, 1812,[65] but Myers had already received a Captain's commission in the regular army five weeks earlier, on March 12, and was fully occupied in his first assignment as a Captain in the *13th United States Infantry.*[66]

5. " 'Sum must spill there blud and others there ink': The Niagara Frontier and St. Lawrence Campaign, 1812-1815."

In March 1812, Myers was ordered by Colonel Peter Philip Schuyler to proceed to Willsboro, New York to establish a chain of recruiting stations throughout Essex County. "....I buckled on my sword," he recalled enthusiastically, "to advance to my station to begin duty as one of the defenders of my country."[67] But his initial enthusiasm was unexpectedly dampened. "....Many of the officers in this District are Farmers who have no idea of punctuality or correctness in accounts," Myers reported to Schuyler, a circumstance that forced him to correct or completely redo the recruiting accounts until late into the long evenings.[68] Despite these complications, Myers succeeded in establishing 13 recruiting stations throughout Essex County. Just a few days after war was declared on June 18th, he was ordered to Plattsburgh where he was obliged to deal with a rash of early desertions.[69] But while there he also trained 330 soldiers so well that they were immediately sent to the military cantonment at Greenbush, New York and received orders from Colonel Isaac Clark "to Take particular Care of the magazine at Plattsburg untill (sic) I can send you further aid —." [70]

Myers then went to the Greenbush Cantonment to join General Henry Dearborn's 4,000 troops for the journey to Buffalo, where they arrived on October 4th after a 16-day march. Over the course of the next six weeks he participated in the capture of the British warship *Caledonia,* the burning of the second vessel, *Detroit,* on Squaw Island (October 9th), General Alexander Smyth's poorly managed attempts to invade Canada (the Battle of Frenchman's Creek, November 28th), and the assaults on the "Red House" and Fort Erie (November 31st).

Myers and his company spent the next seven weeks performing "very severe duty", repeatedly marching the 11 miles back and forth between Buffalo, Black Rock and Williamsville until February 23, 1813, when he was ordered by Lt. Col. Charles Boerstler to move his company to the Williamsville Cantonment.[71]

Once settled at Williamsville, Myers took the time to write to Naphtali Phillips, a close friend and sometime business partner who was also prominent in the Tammany Society and New York's Jewish community. Phillips had recently acquired the *National Advocate,* an influential Democratic-Republican newspaper, and had sent Myers the latest copy. "Sum must spill there blud and others there ink," Myers wrote Phillips with patriotic zeal. "I expect to be amongst the former and I hop you are amongst the latter...." It became clear in the letter that, taken together, Myers's 1811 insolvency and America's declaration of war on Great Britain represented something of a personal reprieve: "It is a fine thing to abandon the persute of welth," he mused. "I never ware hapy in Persute of Riches and now that I have abandoned it I am much more contented." And Myers clearly enjoyed the rewards of command. "....I am considered in a favorable light by my superior officers and Treated with respect by my Equals and Inferiors. I have a Compy that both respect and fear me." There is even a hint of his reading habits, if not something that he acquired in political discussions: ".... A grate man once sayed he would rather be the first in a small viledge than second in rome. It is indead plest to be Premier – in any thing, the responsibility is however grate." [72]

The spring and summer of 1813 were also busy seasons for Myers. He helped arrange medical care for the American wounded following the attack on York (Toronto, April 27) and worked with army engineer Joseph Totten to erect a battery at Snake Hill in preparation for the assault on Fort George (May 27). He was present at the Battle of Stoney Creek (June 6), and would have been at the Battle of Beaver Dams (June 24) had General Boerstler and his 700 men not surrendered prematurely.[73] Myers enjoyed a brief respite from the war during a flirtation with the daughter of a British army surgeon, but also had the melancholy duty of burying 29 American soldiers of his 13th Infantry who had been ambushed and massacred near Fort George. Among the dead was his close friend, Lt. Joseph C. Eldridge.[74]

In the fall of 1813, Myers was on board General James Wilkinson's 328-vessel flotilla as it prepared to make its way toward the St. Lawrence and a final assault on Montreal, but gale force winds and high waves scattered and delayed the flotilla's progress. When it was reported on October 21-22 that 200 sick and wounded soldiers were stranded in two schooners a mile out on Lake Ontario, Myers volunteered to lead the rescue effort. He coordinated 13 trips back and forth with 30 men in 3 Durham boats, saving 160 of the men who were still alive.[75] The delays and turbulent weather had provided the British and Canadian forces on shore and in gunboats to the rear of the Americans ample opportunity to fire on them at will. Wilkinson was forced to halt the flotilla near the farm of John Crysler on the evening of November 10th. There, on the following day, approximately 3000 Americans faced about 1300 British, Canadian and Native forces in the Battle of Crysler's Farm.[76]

As Myers was leading his 86 men onto the battlefield, 63 of them were killed and he was struck in the left shoulder by a musket ball. He had already turned his command over to First Lieutenant William Anderson and was attempting to get back to the American lines when he collapsed from shock and blood loss. The battle was a victory for the outnumbered British forces, but Myers's servant William Williams managed to locate him, get him on a horse and transport him back to the retreating flotilla as it was leaving to get to the Salmon River. Once Wilkinson's army arrived at French Mills (now Fort Covington, New York), Myers, Colonel William Clay Cumming and Colonel James Patton Preston were invited to the home of Dr. Albon Man at Constable for treatment of their severe wounds.[77]

Myers was placed in Man's office on the first floor at the front of the house, and as the doctor was preparing to examine his wound it was noticed that the sun was streaming through the window directly into the captain's eyes. Man's niece, Charlotte Bailey, was asked to stand on a baby's high chair and hang a curtain on the window. The doctor removed 30 pieces of bone splinter during the procedure and found that the musket ball that passed through Myers's shoulder had succeeded in "shattering the head of the humerus, cutting the deltoid muscle destroying the power of the joint and rendering the arm useless, it being also shortened by loss of parts about six inches."[78] Charlotte Bailey, the daughter of Judge William Bailey of Plattsburgh, had been sent to her uncle's Constable home out of the family's concern for her safety, and she would remain there for the next three months helping Myers to recuperate. Man's daughter Susan recalled years later that in the aftermath of his operation, Myers commented on Charlotte's beauty "and the delicate symmetry of her foot and ankle. He said that he fell in love then and there, and it ended in a marriage at my father's house." On the day of the couple's marriage, January 14, 1814, the groom was 37 and the bride 17. [79]

Myers remained at the Man home until March 25th[80] when he rejoined the army at Plattsburgh, now under the command of General Alexander Macomb. Upon his return, he helped quell a mutiny of soldiers (including 2 companies of his own *13th Infantry*) who had not received their back pay, by appealing to their sense of duty, patriotism, and offering them a pointed reminder that they were responsible for upholding their hard-won, honorable reputation.[81] But Myers did not remain long at Plattsburgh. At Macomb's request he rode to Chazy on June 30th with orders to "procure valuable movements of the enemy in that direction and in upper Canada. I went there and frequently rode to St.

Regis and over the line, and I procured much valuable information at the risk of a halter for a neckcloth." He remained at Chazy and continued to go on these fact-finding missions in civilian clothing from mid-August to October, reporting back to Macomb from time to time.[82] Charlotte remained at the Man's, her family still concerned about the British threat to Plattsburgh.

As news of the peace negotiations began to reach America, Myers began to formulate plans for his future. He was hoping for a promotion and wrote to New York State Chancellor James Kent asking him if he would use some of his influence to that end. As Chancellor, Kent held the state's highest judicial position and was the man who just 3 years earlier had declared Myers an insolvent debtor three years earlier, so he may have seemed an odd choice for Myers to ask of such a favor. But Kent was married to Judge Bailey's sister Elizabeth, making the Chancellor Charlotte's uncle. "I can only assure you that if I see any opportunity by which I may be serviceable to you I shall with pleasure embrace it," Kent promised Myers. "From your past Services and sufferings I think you have a just claim to the honorable Notice of the War department." [83] Myers was at Sackett's Harbor the following month when he was ordered to New York City to act as a witness in the court-martial of Colonel Isaac Coles.[84] While in New York he heard from two of his close comrades. Captain Hugh R. Martin expressed his hopes that Myers's name would soon appear on a promotion list[85] and Major Richard Malcolm conveyed his best wishes along with "thoughts of your old messmates." [86] Myers returned to Plattsburgh after the Coles trial and spent part of the third week of February 1815 riding to St. Regis to complete one last mission across the border.[87] Upon his return to Plattsburgh this time he found a letter from Lieutenant James A. Chambers that gives us a brief glimpse of Myers's talents as a host; Chambers recalled with some humor Myers's apparently memorable rendition of a popular barracks lyric regarding "women, war and wine." [88] On March 3, 1815, just two days after Chambers had written his letter to Myers, the U.S. Congress passed an act reorganizing the army, dismissing 5/6 of its officers and reducing the army from 60,000 to 10,000 men,

1/6 of its previous size.[89] Another of Myers's friends, Captain Willoughby Morgan, wrote him to complain with some bitterness how the politicians in Washington now "shake us off –", but saluted his brother-in-arms: "Adieu My friend…. — To Civil life — oh for a good wife –."[90]

Myers's promotion never materialized, but on March 26, 1815, Charlotte Bailey Myers gave birth to the couple's first child, Henrietta, at the Man homestead. Just one day after his discharge from the army on June 15, he was granted a $15 a month disability pension,[91] making his transition back to civilian life somewhat easier. "I then went to New York with about one hundred dollars," he recalled, "to begin the world anew." [92]

6. " '…. Captn Myers is prosperous in business': The Tammany Chieftain as Speculator, Entrepreneur, and Deputy Grand Master, 1816-1828"

Mordecai Myers returned to civilian life during hard times and an unstable economy. Technological advances had stimulated manufacturing capital held by forward-thinking New England entrepreneurs, which led to a boom that soon collapsed. Exports fell and coastal shipping halted, but by the end of the war and the lifting of the last blockades, American food stuffs were being shipped again to European markets.[93] Yet the summer and fall of 1814 had seen the lowest financial downturn in America's history. The national debt rose from $45 million in 1812 to $127 million by the end of 1815. The government sold bonds and treasury notes which were not legal or redeemable in specie. Imports and exports dropped; financial weakness, factional opposition in Congress, disharmony in the administration and a manpower shortage all contributed to the instability.[94] To make matters worse, the government had borrowed $80 million during the war that were now worth only $34 million.[95]

Myers spent his first year as a returning veteran rebuilding his auction business and restoring his old contacts, doing well enough by the end of 1816 to pay $467.26 in duties sales on his auctions.[96] At the same time he entered into a partnership with former Army District Paymaster Joseph Watson helping veterans to obtain or sell their unwanted military bounty lands. That Myers was able to purchase a three-story home on Canal Street and acquire several rental properties attests to the partners' success.[97]

The era of canal construction had already begun by 1817, which meant the development of central and western New York trade to New York City, along with a growing agricultural and manufacturing period during which there was a rise in both population and prosperity. In 1812 alone, 44% of all imports sold in New York City were sold through auctions.[98] With increasing profits through his auction business, Myers invested more heavily in city real estate. "I am informed you are going in your usual systematic way, that you live comfortably and respectfully and that notwithstanding the nosiness of the time, you are accumulating property –...", Judge Bailey wrote to his son-in-law. "These circumstances gives (sic) me & my family great Satisfaction –. "[99] Charlotte's brother Theodorus visited the Myers in 1823 and assured his father that "Your Daughter Charlotte & family are well — and Captn Myers is prosperous in business."[100]

Myers did indeed prosper beyond his expectations over the next few years. In 1824 the Grand Lodge of the State of New York appointed him Chairman of the Committee of Arrangements to welcome and honor the Marquis de Lafayette when the aging Revolutionary War hero arrived in New York on August 16 to begin his triumphal tour of America.[101] In that same year, Myers had invested in and become secretary of the "Florida Association of New York". Three years earlier, he and several investors had purchased 30,370 acres of the Alachua Tract in East Florida from Fernando and Jose de la Maza of Cuba. The Association had already transported families, food, supplies, livestock and building materials to the Tract for settlement and development when questions about the validity of the original Spanish grants under the Florida Cession of 1819 were raised. Myers and the members of the Association testified at New York City Hall on October 30, 1824 in an effort to resolve the issue, which was concluded in the Association's favor by the U.S. Supreme

Court in 1832.[102] He continued in those years to expand his local business into a "Military and General Agency office", remitting taxes to Missouri, Illinois, Ohio and Arkansas for lands held there, and purchasing, selling or exchanging property in those states for customers.[103]

On January 8, 1826, the anniversary of the Battle of New Orleans, several officers of America's "second war of Independence" met at Myers's home to draw up a petition to Congress requesting "a grant of public lands" for "their services, losses and suffering….according to rank and former practice." Myers, who served as secretary for the group, was joined that evening by Majors George Howard and Clarkson Crolius, and Colonels Joseph Watson, Joseph Lee Smith, Gilbert Christian Russell, and James R. Mullany, who called themselves the "Military Society of the War of 1812." [104] The Society became part of a larger national movement calling for greater compensation for veterans of the War of 1812 and eventually merged with the New York State Veteran Corps of Artillery (VCA), founded in 1790 by New York City veterans of the American Revolution. Myers was appointed a VCA Brigade Major in 1825 and a VCA Adjutant in 1826, after which he was often referred to as "Major Myers" or simply "the Major".[105] He was also at this time raised by the Tammany Society to the position of "Chief" or "Chieftain", which gave him the authority to recommend proven Democrats for local political appointments. [106]

On September 11, 1826, a hard-drinking 50-year-old bricklayer named William Morgan was arrested and jailed for theft at Canandaigua, New York. Morgan was released, arrested again and jailed on the night of September 12th for failing to pay a debt of $2.69. Someone paid for his release that same night, but he was immediately kidnapped and taken away in a yellow carriage, first to Rochester, then Fort Niagara, across to Canada, then back to Fort Niagara, after which he disappeared. Morgan was a Freemason who had been denied entrance into a new Masonic lodge at Batavia, New York. Angry and humiliated, he published a book entitled *Illustrations of Masonry* which revealed many if not all of the Fraternity's secrets. His disappearance and the subsequent controversy it generated, "the Morgan Affair" or "the Morgan Excitement", caused suspicion to be cast upon all Free-masons and led to the rise of a new Antimasonic Party. Antimasonic conventions began to meet in 1827, along with the appearance of Antimasonic books, newspapers, clubs and lectures. Masons were stoned and clubbed, their businesses boycotted, and they were barred from serving on juries. Arrests, trials and indictments for the perpetrators of Morgan's abduction lasted until 1831, though the political and social impact of the events were felt for another six years. [107] Between 1827 and 1836 the number of New York State Lodges dropped from 480 to 70. Mordecai Myers was the Deputy Grand Master of the Grand Lodge of New York State throughout the entire Morgan period and spent much of his time and energy providing moral, political and even financial aid and advice to beleaguered Masons and their lodges throughout the state. Myers's efforts on behalf of Freemasonry were never forgotten; even after his death decades later, he was still being hailed as a hero of Freemasonry at the State Lodge's Grand Encampments held annually in New York City. [108]

The auction system that had flourished after the War of 1812 and provided Myers with much of his income had begun to decline by the late 1820's. There were more American domestic importers who had established their own networks in Europe as well as new transatlantic banking arrangements and commercial credit expansion; in time the old auction system "could no longer compete with the newer distribution mechanisms of the economy" and it gradually became obsolete. [109]

Fortunately, the political landscape had changed as well during those years. Martin Van Buren's political machine, the "Albany Regency", had been exercising control over virtually all state offices from the Governor to local postmasters since the early 1820's. [110] Mordecai Myers's distinguished military record, his reputation as a principled businessman, and longstanding dedication to the Democratic Party had made him a valued asset to the Regency. As a 32nd Degree Freemason he also appeared willing and able to withstand and oppose any assaults leveled at the Masons originating from the New York State Legislature's growing Antimasonic faction and its statewide supporters. On October 2, 1828, Myers received the Democratic nomination to represent New York City's 8th Ward in the State Assembly. After a month of intense campaigning Myers won the seat and became the first Jew to serve in the New York State Legislature. [111]

7. "'…. a Jewish law-giver in a Christian land' – The Assembly Years, 1828-1834"

Myers wasted no time applying himself to Assembly business. In that first year alone, he brokered a trade agreement between a consortium of New York City investors and Cuban commodities traders, helped obtain relief for the widow and heirs of Revolutionary War veteran Isaiah Wool, and saw to the incorporation of both the Manhattan Gas Light Company and the City Coal Association.[112] Re-elected in 1829, he supported all of the then-popular legislation promoting internal improvements, including the construction of bridges, canals, dams, turnpikes and schools, in addition to earmarking aid for the aged, orphaned and indigent.[113] Myers chose not to run in 1830. The New York Common Council had granted the Manhattan Gas Light Company a lucrative franchise covering Manhattan above Grand and Canal Streets that year and Myers was selected as one of the Company's directors on February 26, 1830. Since he was the Assembly's sponsor of the Company's incorporation just two years earlier, he was probably attempting to avoid any hint of collusion or conflict of interest. [114]

Myers commenced his third term in the Assembly in January 1831 as a member of a 5-man "Committee on the Militia and Subjects relating to the Public Defense".[115] As chairman of that same committee the following year, Myers pressed for the expansion of the state's pension provisions for the benefit of the state's Revolutionary War veterans to also be applied to veterans of the War of 1812. Myers and his Democratic colleagues also petitioned their state's Congressmen "to use their exertions to procure for the officers of the Late war, a gratuity of lands, as a reward for their losses, sufferings and privations during the second war of independence." The request received the support of both the Assembly and the Governor, but another 25 years would pass before the nation acted on this and similar requests from most of the states. However, Myers did successfully push through a reduction in the number of militia parades on the grounds that they were "detrimental to morale and took laborers away from their places of useful employment unnecessarily." [116]

In 1831 the Assembly was presented with 26 petitions from around New York State protesting the payment of legislative chaplains who performed prayer services before each session. The petitions were referred to a committee composed of David Moulton (Oneida County), John C. Kemble (Rensselaer County) and Mordecai Myers (New York City). In January 1832 the legislature motioned to invite the Albany clergy to say opening prayers in rotation. Moulton objected, but the chaplains were hired and paid as usual for that session. [117]

The "Moulton-Myers Report" was presented to the Assembly on April 16, 1832. It states in part: "Despite the numerical predominance of Christians, christianity as such is not the law of the land. It is the constitution which is supreme. By the constitution, all religions – the mosque, the synagogue, the Christian church, and all other churches – stand on equal footing....The whole question of appointment and payment of legislative chaplains is an unauthorized extension of the constitution. Whatever statutes allow legislative chaplaincies are unconstitutional and should be repealed." [118]

The Report had its opponents. James Willson, pastor of Albany's Presbyterian Church, published a pamphlet, *Prince Messiah's Claims to Dominion over All Governments; And the Disregard of His Authority by the United States, in the Federal Constitution,* in which he insisted that it was the duty of all American citizens "to obey Christ...." Willson denounced the Founding Fathers as non-believers and infidels who supported the "godlessness of the Constitution", which he called a document that allows "Atheists, Deists, Jews, Pagans" and other non-Christians to hold office. [119] It happened that Myers was serving that year as the New York State Grand Lodge's Deputy High Priest [120]

This, along with Myers being the only Jew in the Assembly and thus even more conspicuous, made him Willson's particular target. "In the discussion, Mr. Myers, a Jewish representative from New York , took an active part against prayer -," Willson pointed out. "This was in character, as all the chaplains pray in the name of Jesus of Nazareth, whom his fathers crucified, and whom he rejects as an imposter. Yet for a Jew to oppose prayer, as offered up in the legislature of a Christian country, was an act of discourtesy to the Christian people whom he represents." [121] This was the first and only time in his political life that his ethnic background emerged in connection with a public issue. But in what must have been a relief for Myers and the Democratic Party, the Assembly voted 95 to 2 "to delete Willson's name from the list of rotating Albany clergy who offered up prayer at the beginning of each day's session." Willson himself was burned in effigy in front of the Assembly's entrance and his pamphlet burned in a public bonfire nearby,[122] more for his disparaging remarks about America's Founding Fathers than for the allegations of deicide he levelled at Myers's antecedants.

The 1833 legislative session passed uneventfully for Myers, who served on committees examining trade and manufactures and the penitentiary system. [123] The next year, Myers's sixth and final one in the Assembly, he served on the Ways and Means Committee and saw through the incorporation of the Schenectady Savings Bank. [124] But the issue of legislative chaplaincies arose one more time in 1834. Solomon Southwick, an editor and one-time Freemason, joined the ranks of the Antimasons in a bid to gain political office. Using the pseudonym "Sherlock", Southwick wrote and published a series of letters under the title, A Layman's Apology For The Appointment of Clerical Chaplains in which he charged that the Moulton-Myers Report "aimed a vital blow at the Christian Religion." [125] As for the authors of the Report: "The first named is an avowed Infidel; the second is a Jew;" [126] Southwick echoed James Willson's claim that Myers did not sufficiently appreciate having been elected to his position, "a Jewish law-giver in a Christian land – a fact never before heard of in any other Christian land – was it grateful or magnanimous in him to aim a blow at the Ministers of Christianity?" [127] Southwick's efforts fell on deaf ears and he soon after fell into political obscurity and personal bankruptcy.

It is not known if the controversy over legislative chaplaincies was a factor in Myers's decision not to run again for the Assembly, but at least part of that decision may have been a wish on his part to retire. He was "nominated assistant alderman, and once alderman of the eighth ward" of New York City after leaving the Assembly, but "declined both nominations." [128] His investments and city rental properties were generating a steady income, and he was anticipating the same from his holdings in Florida, upstate New York and in the Western Territories. Now nearing 60, Myers was firmly established and well able to provide his large family with all of the advantages appropriate to their station.

Myers may also have been anxious to move his family out of New York City. The city had become more congested and unsanitary than ever; memories of the Asiatic cholera epidemic that took 3000 lives in the city alone from July 4 to October 1st, 1832, were still fresh. Social and political unrest were becoming more common. The New York City municipal elections of August 1834 led to street brawls and riots. Poverty-stricken Irish immigrants facing prejudice by "Nativist" Americans formed street gangs, resulting in the "Five Points" riot of June 21, 1835. And on the evening of December 16, 1835, the "Great Fire" of New York destroyed 674 buildings in 19 blocks of the city's commercial and residential areas, causing $20 million in damage.

Myers decided to move 130 miles up the Hudson River to the Columbia county village of Kinderhook. From there he could manage his properties and raise his children in a quieter, safer rural setting. The village held another attraction for Myers the loyal, lifelong Democrat; with Vice-President Martin Van Buren succeeding Andrew Jackson to the Presidency, it was not lost on Myers that Kinderhook was the new President's native town....the term "OK" ("Old Kinderhook") was sometimes used to describe Van Buren .... and thus a vital link to the still-powerful Albany Regency.

8. "Kinderhook, the Panic of 1837, and the Return to New York City, 1835-1848"

Myers purchased a three-story Federal style home on 11 acres in the center of Kinderhook from Judge Aaron Vanderpoel, settled his family there by the summer of 1835[129] and quickly took his place as one of Kinderhook's leading citizens. He received an appointment to a citizens' committee charged with forming a local militia to reinforce the brigade stationed at his old Greenbush Cantonment just 15 miles away. The appointment came from General Henry James Genet, who had served in the Assembly with Myers in 1832 and recently succeeded Major General Stephen Van Rensselaer as Commandant of the State Militia.[130] Genet also invited Myers to march in an upcoming Independence Day procession in Troy, New York. "I have Greate Plesure in accepting your appointment," Myers replied, "and will, if my health Permits Join you and your Command at the City of Troy at an Early Hour on the 4th of July, the name of which day causes the Blude of the old Soldier to cours Swiftly through to and from the fountin of an American Heart—" [131]

Myers was also named to the Board of Directors of the newly-established Kinderhook Bank and a trustee of the prestigious Kinderhook Academy. [132] His eldest sons William and Theodorus studied law with the Kinderhook firm of Vanderpoel and Tobey, and accompanied their father to the Columbia County Democratic Conventions. The younger Myers children — Henrietta, Frances, Louisa, Charlotte, Katerine, Edward, Charles and Sydney — attended school, enjoyed rounds of parties and games, sleigh rides in the winter, and Sunday night teas hosted by their parents. Mordecai and Charlotte set aside some evenings for a carriage ride, and the Major played the congenial host to a constant flow of friends, relatives and political associates.[133] Though officially retired, Myers never refused requests from any veteran of the War of 1812 for help or advice. Henry Thomas, a shoemaker who once served in the *13th Infantry*, wrote asking for advice in obtaining his bounty lands. Myers assured Thomas that he recalled him and added that "it alwase gives me Greate

Plesure if I Can be of the [b]est Service to anny of the Brave Soldieirs of the late [war] more Especialy to Such as served under me —" [134]

1839 would prove to be the happiest of the Myers' Kinderhook years. Mordecai was elected to a one-year term as "President" (mayor) of the village on May 9, 1839.[135] In that capacity he publicly welcomed Martin Van Buren when the sitting President visited his hometown. [136] And on September 3rd, the Myers' eldest daughter Henrietta married Kinderhook merchant Peter S. Hoes at the Myers' home. [137] Hoes was the nephew of Martin Van Buren, so the marriage now cemented the Major's Democratic political associations into familial ties.

Unfortunately, the Myers' idyllic life in Kinderhook proved to be short-lived. In July 1840 the family received word that their 23-year-old son William had died unexpectedly of fever in the town of Farmington, Illinois.[138] William, recently been admitted to the Columbia County bar, had been in Farmington, Illinois, overseeing a land transaction for his father and was being groomed to assume control of his affairs, so his death was an enormous blow to the family. "Mills" as he was called by family and friends, had been cared for by a Mr. and Mrs. William Hatch of Farmington, and a few days after hearing of his son's death, Mordecai wrote an eloquent and heartfelt letter to Mrs. Hatch:

Dear Madam

    The first Paraxisms of grieffe Having Pased off and Calm resan Having resumed its Empire I Imbrase this time to acknalge the Kindnes of yoursalfe and family toards my Son During his Illnes at your Hause and for you attention and that of your family and frends in Having His Mortil remains Depasited in our Mother Earth in a Disant and Becaming Maner Will you please Make My Worm acknalagments to Evry Individual in your Village that showed hime kindness Whilst Living or respect when Died it apers to me and my family that you have all By acts of kindness Indired yoursalves to us.....But it Has Plesed God to reclaim His Immortable Spirit and it is our Duty to acques — without murmering at His will your

*Slite acquaintance with his amenable qualitys must con-
vince you How Dear he was to His 9 sisters and Brother
and family connectians May God Give rest to his Sole*

I am Madam
your frend and Most obedt
Mord[e]cai Myers

Myers then gave detailed instructions for the disposition of William's personal effects and the carving and delivery of a headstone for his burial site, which to this day remains intact and legible in Farmington's Oak Ridge Cemetery. [139]

Mordecai had already been struggling financially when William died. The proliferation of banks, the issuance of bank notes and loans, widespread state debts due to excessive borrowing for internal improvements, and over-speculation in western lands all had disastrous consequences for the nation's economy. Andrew Jackson's July, 1836 "Specie Circular", requiring all payments for public land be paid in gold or silver coins, made the situation worse. Banks closed their doors, strikes became commonplace, and there were even fears of civil war. The resulting Panic of 1837 plunged the nation into a depression that lasted for six years, the worst until 1929. Most of the tenants who occupied Myers's heavily mortgaged properties were now unable to pay their rents. Myers attempted but failed to sell many of his holdings; anticipating the worse, he began to make arrangements to sell his Kinderhook home. [140]

While dealing with his finances, Myers made an attempt to augment his shrinking income by applying for the recently-vacated post of Commissary (Inspector) General of Military Stores for New York State and sent nearly identical letters to Major General Winfield Scott and former New York City Mayor and Tammany leader Isaac L. Varian asking for their support. He reminded Scott, whom he referred to as one of "a few of My old Militarry frends" of his services on the Niagara Frontier, while emphasizing to Varian his lifelong "adherence to Republican principles", to say nothing of the money and lobbying efforts he had contributed to the Democratic cause over the years. Myers also contacted one of the Albany Regency's major figures, Silas Wright, but the power and influence of the Regency had begun to fade when Van Buren failed to be re-elected in 1840, and Myers's efforts were unsuccessful. [141]

The Kinderhook property was finally sold in 1843 and Myers moved the family back to New York City. His brokerage, loan and land business had begun to show gradual signs of improvement during the next two years when the family suffered two more tragic losses. On November 7, 1845, the Myers' 19-year-old daughter Maria Louisa died suddenly of unknown causes. [142] And Charlotte Bailey Myers, Mordecai's beloved wife of 34 years, had contracted consumption (tuberculosis) while in New York and died on February 15, 1848, at the age of 52. [143]

Mordecai decided to move again, this time to Schenectady, New York, located 16 miles from Albany in the Mohawk Valley. He had become familiar with the city while serving in the Assembly, and he and Charlotte had once considered it "a pleasant abode for their declining years." [144] The New York City home was sold and in 1848 the Myers family moved to Schenectady, where the Major would spend the last 23 years of his long life.

9. "'....Whear I am So whel known and Pret Genrally respected –': Schenectady, New York, 1848-1871"

When Captain Mordecai Myers of the *13th U.S. Infantry* rode with General Dearborn's army through Schenectady in the fall of 1812 on the way to the Niagara Frontier, it was already a substantial city of nearly 6000 people. [145] The Erie Canal, completed in 1825, had transformed "the Gateway to the West," as Schenectady was known, into one of the busiest commercial hubs of the Eastern seaboard, now with a population of 9000. The Schenectady Locomotive Engineering Manufactory, the Westinghouse Agricultural Works, and dozens of factories producing brooms, carriages, stoves, candles, soap and other essentials employed hundreds of German and Irish immigrants. Union College, under the leadership of Eliphalet Nott, had become a leading academic institution.

Myers had found a comfortable and spacious house at 65 Centre Street (now Broadway) where he settled his still-large family.[146] At 74, he still had five children at home: Katherine ("Kate", 24); Charlotte ("Lottie", 22); and 3 still in their teens: Edward ("Eddie", 17); Charles ("Charlie", 15, who took "William" as his middle name in memory of his beloved eldest brother), and Francis ("Fannie", 13), the youngest and only child born at the family home at Kinderhook. [147] Henrietta Myers Hoes still lived in Kinderhook with her husband and two sons; Theodorus practiced law in New York City and Sidney, recently married, was a bookseller and stationer at Galesburg, Illinois. In addition to his close-knit family, Myers also had two close friends from the *13th Infantry* living nearby. Hugh R. Martin, a Schenectady native who had served with Myers on the Niagara Frontier, operated a large brewery in town,[148] and John Keyes Paige, who was with Myers at Chrysler's Farm, was a successful attorney in the city.[149] His statewide reputation as an ardent defender of the Lodge during the "Morgan Excitement" was well known among local Masons, and many of the city's leading businessmen recalled with approval former State Assemblyman Myers's support for the incorporation of the Schenectady Savings Bank.

But Mordecai was never content to settle into domesticity or remain politically inactive for very long. In September of 1850 he was elected chairman of the Schenectady county Democratic Convention, then delegate to the State Congressional Convention. [150] Barely six months after serving in that capacity he received the Democratic nomination for mayor of Schenectady and was elected to a one-year term. He spent that year quietly overseeing the city's business before turning the office over to his successor, A.A.Van Voast,[151] then served as one of four delegates to the New York State Senatorial Convention of August 27, 1853.[152] In that same year he was elevated to the office of Grand Master of the Grand Lodge of the State of New York (a post he held through 1856),[153] and began to write the letters to his son Theodorus which were published posthumously a half-century later as his *Reminiscenses.* [154]

Myers was nominated for a second term as mayor of Schenectady on March 29, 1854, and won again, but now faced the 10-member Common Council dominated by 8 Whigs, who commenced blaming the recently elected Democrat for the city's financial troubles. [155] Myers was used to this type of political wrangling, but he soon faced a statewide issue that could not be avoided. In April, 1854 New York Governor Horatio Seymour signed a bill authorizing the establishment of a free schools system throughout the state. The Schenectady Common Council was given the power to raise the money for that purpose by imposing taxes on real and personal property. Myers opposed the new Board of Education's tax assessment and the Council's power to implement it, calling it "a burden too atrocious to be borne." He resigned his office in protest on March 21, 1855, just two weeks before it would have ended. City Alderman Casper F. Hoag served Myers's unexpired term until April 5th. [156]

Myers passed the next few years attending annual Masonic Encampments in New York and making speeches for visiting politicians. While vacationing with his children at the popular resort of Richfield Springs, New York, he founded Richfield Springs Masonic Lodge No. 482, which is still active today. [157] On October 19, 1860, a 3-man Democratic committee nominated the 84-year-old to represent the 18th Congressional District (which included Schenectady County) in a bid for a seat in the United States Congress. [158] Theodorus, increasingly worried about his father's age and health, expressed his concerns, but Mordecai ignored him. "I Have Never refused to Serve my Country in a Good Cause," he explained in a letter. "I accepted and Now Stand firm....Your father Will Retain his Characture as an Honest Politiccan...." [159] The nomination was announced in the press, but Myers decided to withdraw from the race without explanation.

In March 1861, historian Benson J. Lossing visited Schenectady while researching his projected history of the War of 1812. It is not clear whether the well-travelled Lossing met and interviewed Mordecai Myers, or simply asked him for a written account of his service during the war, but when Lossing's

Pictorial Field-Book of the War of 1812 was finally published in 1868 it included a description of Myers's role in the October 1813 rescue of the 160 soldiers on Lake Ontario. Lossing cited "the now venerable Mordecai Myers of Schenectady, N.Y., to whom I am indebted for an interesting narrative of the events of this campaign" and included a brief sketch of Myers's life, as well as a facsimile of his distinctive signature. [160] Myers also provided his neighbors with a colorful glimpse of his more chivalrous past. Schenectady was experiencing a minor crime wave during the 1860's, with burglaries, pickpocketing and public intoxication becoming more common. The usually dignified octogenarian was once seen standing guard at his front door, wearing a red night cap and eyes blazing as he brandished the saber he carried during the war, threatening a would-be burglar seen lurking near his Centre Street home. [161]

Myers had once written that he had "never suffered misfortune to dampen my energies, nor prosperity to elate me unreasonably" [162] but his proverbial calm in the face of hardship was tested still again when he lost his two youngest sons. 31-year-old Eddie died at the Centre Street home on July 15, 1863;[163] Charlie, just 28, also died at home three months later. [164] Several months later Mordecai wrote to his daughter Charlotte, now married with two daughters and living in Yonkers, thanking her for an invitation to meet her in Albany, then move in with her family. He thanked her but concluded that his "advanced age and numerous Infirmaties, although Not Imidiatly dangerous yeate Verry tormenting" prevented such a move. "I Have concluded that My life is Here at Home....And So Near our family resting Place and here whear I am so whel known and Pret Generally respected — and do not think that I Have Gote the Blues or that I Look upon old age and approaching Death with feer and alarm Nothing of the kind I assure you I feel quite redy to Retire from *this* Ceracy *World* When Ever it Shall Pleas God to Gate me and in full hops of a Better one." [165]

Myers sold the Centre Street home at public auction a little more than a year later, [166] and boarded for one year at the fashionable Carley House, two blocks away, until moving in with his daughter Fannie and her husband Edgar M. Jenkins. [167]   On August 16, 1869, Myers purchased the house at 71 Union Street next door to the Jenkins, where he and daughter Kate would reside for the rest of both their lives.[168]   At 93 he now stayed closer than ever to home, welcoming visits from his children, grandchildren and friends, maintaining his congeniality while regaling those present with stories from the war. One of his grandchildren left a vivid but regrettably brief recollection of the Major from that period: "His carriage, until old age, was noticeably erect, his bearing both dignified and martial, his manners most polished, his conversation elegant and always interesting, his presence commanding —" [169] He also lived long enough to receive a copy of Lossing's *Field- Book of the War of 1812* as a gift from Theodorus. "I thank you Very mutch for the Book you sent me the auther Has Erected 2 Monuments to my Memory...." He was still somewhat bitter even after 56 years that the government "disbanded me with the por Pay of a Captn at the Peace", but added with obvious satisfaction that "I am Plesed to find there is 2 hist notes to my Credit...." [170]

A year later, on the morning of January 20, 1871, Mordecai Myers died quietly in his sleep, just 4 months short of his 96thbirthday. The Schenectady *Evening Star* reported that the Major had enjoyed a "social smoke" with an old friend the night before, and described at length the highlights of Myers's early life, military exploits and his political and fraternal associations. [171] According to local tradition, Myers had summoned a local grocer named Alexander Susholz to his home shortly before his death, requested that his remains be prepared according to Jewish custom and that he be interred in the local Jewish cemetery, though Myers had purchased a family plot in Vale Cemetery in 1857. Susholz (1806-1883) was a German-Jewish immigrant who had served 5 years in the Prussian army and arrived in Schenectady as a poor peddler in 1848, the same year as Myers's arrival. He was also a founder of Schenectady's first synagogue, Congregation Gates of Heaven, and Myers may well have been experiencing some need to reconnect with the Jewish community. But his surviving children, all of whom were raised as Episcopalians, saw to it that their father's funeral was carried out by the pastor of St. George's Episcopal Church and the Freemasons, and that he was interred at Vale Cemetery. [172]

Myers's last will and testament reflected his lifelong links to his fraternal and religious associations, as well as his sense of civic responsibility. He left bequests of $200 each to five Jewish organizations, including schools, hospitals and orphan asylums, and $250 for "the North American Relief Society for the poor of Palestine, to be applied for the benefit of the poor Israelites of Palestine." Another $200 went to the Grand Lodge of the State of New York, and $400 to the City of Schenectady toward the eventual purchase of a site for a city hospital.[173] The *Evening Star* noted at the time of his death that while the Major's religion had indeed "differed from that of many around him", it was yet "inferior to none in security and devotion." But it was the Reverend Dr. William Payne of St. George's Episcopal Church who may have best described Mordecai Myers's life and character when he stood before the Major's family, friends and associates and quoted Antony's words from Shakespeare's *Julius Caesar*:

> "His life was gentle; and the elements so mixed in him, that nature might stand up and say to all the world, *This was a man*." [174]

One last inscription on Myers's monument in Vale Cemetery reads:

> "The Historic pen hath writ his spotless name Among the good and brave, the men of deathless fame."

It was a fitting and unadorned tribute to a man who never ceased thinking of himself as "the old Soldier", an honored veteran of America's "Second War of Independence", and one of the last surviving officers of the *13th United States Infantry* who had fought at Crysler's Farm.

Notes on Part One:
The Life & Times of
Mordecai Myers

(1) *Schenectady Evening Star.* January 20, 1871, p. 3; January 21, 1871, p. 3; January 25, 1871, p. 3; *Schenectady Daily Union,* January 24, 1871, p. 3; January 25, 1871, p. 3.

(2) Mordecai Myers, *Reminiscenses 1780 to 1814 – Including Incidents In The War of 1812-14. Letters Pertaining To His Early Life* (Washington, D.C.: The Crane Company, 1900). Hereafter cited as *Reminscenses.*

(3) "Meir", *Encyclopedia Judaica* (Jerusalem, Israel: Keter Publishing House Jerusalem Ltd., 1972), XI, pp. 1240-42. Hereafter cited as *EJ;* Dan Rottenberg, *Finding Our Fathers: A Guidebook to Jewish Genealogy* (New York: Random House, 1977), p. 290.

(4) Steven Runciman, *A History of the Crusades* (London: The Folio Society, 1994), Vol. 1, 111-117; Emil Lengyel, *1,000 Years of Hungary* (New York: The John Day Company, 1958), pp. 33-36; A. Schreiber, *Jewish Inscriptions in Hungary From the 3rd Century to 1686* (Budapest and Leiden, 1983), pp. 314-318.

(5) Raphael Patai, *The Jews of Hungary: History, Culture, Psychology* (Detroit: Wayne State University Press, 1996), pp. 125-52, pp. 190-91, p. 204.

(6) "Familiants Laws", *EJ,* VI, pp. 1162-64; Istvan Veghazi, "The Role of Jews in the Economic Life of Hungary" in Randolph L. Braham, ed., *Hungarian Jewish Studies* (New York: World Federation of Hungarian Jews, 1966,1969), Vol. II, 50-53. Hereafter cited as Braham; Erno Marton, "The Family Tree of Hungarian Jewry", in Braham, Vol. I, pp. 32-49; In order to tax the Jews in his realm more efficiently, Charles VI ordered the first comprehensive census of Hungarian Jewry, the "Conscriptio Judeaorum" of 1735-38 and updated by Maria Theresa in 1752-54. The census indicated that of the 11,621 Hungarian Jews, 2826 of Austro-German descent lived in Moson and Sopron counties, directly on the Hungarian-Austrian border. The family name "Myers" and its variants – "Meyers", "Mayers" or "Majers" – appear most frequently in those counties and represent the likely origins of Mordecai Myers' father's family. Material specific to the Myers family was researched for the editor courtesy of Dr. Kollega Tarsoly Istvan of the Magyar Orszagos Leveltar (MOL), Budapest Hungary in the Helytartotanacsi Leveltar (HL) -C29 – Acta Judeorum/Conscriptio

Judaeorum, 1725-28 and MOL-HL-Acta Judaeorum/C-29/-No.1-153-No.4.-1752-1754.

(7) C.A. Macartney, *Maria Theresa and the House of Austria* (Mystic, Connecticut: Lawrence Verry Inc., 1969), p. 19; "Maria Theresa", *EJ*, VIII, p. 991.

(8) Benjajmin Myers died on November 20, 1776 at the age of 43. See "Items Relating to the Jews of Newport", *Publications of the American Jewish Historical Society*, Vol. XXVII (1920), p. 198 and "Items-Seixas Family", p. 350. Cited hereafter as *PAJHS*; Rachel Myers died on March 30, 1801 at the age of 56. See David De Sola Pool, *Portraits Etched in Stone – Early Jewish Settlers, 1682-1831* (New York: Columbia University Press, 1952), p. 285. Hereafter cited as De Sola Pool, *Portraits*.

(9) Charlotte Baum, Paula Hyman, & Sonya Michel, *The Jewish Woman in America* (New York: The Dial Press, 1975), pp. 4-11; Anita Libman Lebeson, *Recall to Life: The Jewish Woman in America* (New Jersey: Thomas Yoseloff, 1970), pp. 31-35, pp. 47-56, pp. 89-91.

(10) *Reminiscenses.*

(11). Charles Elmer Allison, *The History of Yonkers* (New York: W.B. Ketcham, 1896), pp. 55-82.

(12) "Registry of circumcisions by Abm. I. Abrahams From June 1756 to January, 1783 in New York, in Hebrew and English" in "Misc. Items, N.Y. Congregations", *PAJHS*, XXVII, 150-151.

(13) Morris A. Gutstein, *The Story of the Jews of Newport: Two And A Half Centuries of Judaism, 1658-1908* (New York: Bloch Publishing Co., 1936), pp. 157-164. Hereafter cited as Gutstein; Sheila Skemp, *A Social and Cultural History of Newport, Rhode Island 1720-1765* (Ann Arbor, Michigan: Xerox University Microfilms, 1974), pp. 217-250, pp. 271-272, pp. 342-371. Hereafter cited as Skemp; Stanley F. Chyet, *Lopez of Newport: Colonial American Merchant Prince* (Detroit: Wayne State University Press, 1970), Et passim.

(14) Holly Snyder, "Reconstructing the Lives of Newport's Hidden Jews, 1740-1790", in George M. Goodwin and Ellen Smith, eds., *The Jews of Rhode Island* (Waltham, Massachusetts: Brandeis University Press, 2004) pp. 27-39. Hereafter cited as Snyder.

(15) "Items Relating to the Jews of Newport", *PAJHS*, XXVII, pp. 212-214; Malcolm Stern, "Myer Benjamin and his Descendants: A Study in Bio-graphical Method", in *Rhode Island Jewish Historical Notes* (Providence, Rhode Island: Rhode Island Jewish Historical Association), Vol. 5, Number 2: 1968, p. 143, n. 18. Hereafter cited as Stern; George Champlin Mason,

Reminiscenses of Newport (Newport, R.I.: Charles E. Hammett, Jr., 1884), p. 61. The British forces at Newport were forced to tear down approximately 200 residences, warehouses and outbuildings to obtain firewood during the harsh winters of their occupation. The Myers' residence was likely a casualty of these demolitions. St. Paul's Methodist Church, erected in 1806, now occupies the site on Marlborough Street.

(16) S. Broches, *Jews in New England: Six Historical Monographs* (New York: Bloch Publishing Co., 1942), p. 381; Stern, 137b.

(17) Morris U. Schappes, ed., *A Documentary History of the Jews in the United States 1654-1875* (New York: Schocken Books, 1971), pp. 35-37. Hereafter cited as Schappes.

(18) Stern, pp. 134-35, p. 143, n.3.

(19) Stern, p. 143, n.22

(20. Gutstein, p. 114.

(21) Jacob R. Marcus, *The Colonial American Jew, 1492-1776* (Detroit, Michigan: Wayne State University Press, 1970), Vol. II, 1062-1063; *United States Jewry, 1776-1985* (Detroit: Wayne State University Press, 1989), Vol. I, pp. 251-262;Vol. II, p. 82. Hereafter cited as *U.S. Jewry*.

(22) Snyder, p. 34.

(23) Stern, p. 141, p. 143, n.21

(24) *PAJHS*, XXVII, p. 214; Richard Polenberg, *The World of Benjamin Cardozo: Personal Values and the Judicial Process* (Cambridge, MA: Harvard University Press, 1997), p. 14.

(25) *PAJHS*, XXVII, 198, 350; De Sola Pool, *Portraits*, p. 285.

(26) Neil B. Yetwin, "American-Jewish Loyalists: The Myers Family", *The Loyalist Gazette* (Toronto: the United Empire Loyalists' Association of Canada), Vol. XXVIII, No. 2, Fall,1990, p. 14; Abraham Myers served in the *1st, 2nd* and *3rd New Jersey Volunteers*. See Vol. 1851, p. 52, Reel C-3873; Vol. 1856, p. 83, p. 96, p. 110, Reel C-4216; Vol. 1855, p. 43, p. 58, p. 66, Reel C-3874; Vol. 1855, p. 86, Reel C-3874; Photocopies courtesy of the National Archives of Canada; Myers Solis-Cohen, *The Hays Family in America* (1931, unpublished typed manuscript courtesy of the late Alva Middleton, the last descendant of Benjamin Myers); Salomon Solis-Cohen, "Note concerning David Hays and Reuben Etting, their Brothers, Patriots of the Revolution", *PAJHS*, II (1894), pp. 63-72; William S. Stryker, "The New Jersey Volunteers — Loyalists — In The Revolutionary War", in *The Capture of the Block House at Toms River, New Jersey, March 24, 1782* (Trenton, New Jersey, 1883), pp. 3-67.

(27) Christopher Moore, *The Loyalists: Revolution, Exile, Settlement* (Toronto: Macmillan of Canada, 1984), p. 93.

(28) Rachel Myers to Sir Henry Clinton. New York, April 3, 1781. Photocopy courtesy of the College of William and Mary: The Colonial Williamsburg Foundation. And edited version of this petition, omitting the names of the officers of the *43rd* and *23rd British Regiments* who signed it in support of Rachel's claims and recommending her "as an object of Charity from the Largeness of her Family" was published in Jacob R. Marcus, *American Jewry – Documents – Eighteenth Century* (Cincinnati, Ohio: The Hebrew Union College Press, 1959), pp. 273-274.

(29) Esther Clark Wright, *The Loyalists of New Brunswick* (Fredericton, New Brunswick, Canada: 1955), pp. 46-50; D.G. Bell, *Early Loyalist Saint John: The Origin of New Brunswick Politics 1783-1786* (Fredericton, New Brunswick: New Ireland Press, 1983), p. 17, p. 20.

(30) William Russell, "The Landing of the United Empire Loyalists in New Brunswick", *Miscellaneous Research Papers, Manuscript Report Series, No. 216* (Ottawa: Parks Canada, 1975-77), p. 151-159.

(31) RG 20. Series A, Vol. 12 (1784) #43. Courtesy Public Archives of Nova Scotia.

(32) "The Petition of Rachel Myers to His Excellency Thomas Carleton". Parr: January 24, 1785. RS-108, Public Archives of New Brunswick, Fredericton, New Brunswick, Canada. Hereafter cited as PANB.

(33) Rachel Myers to Governor Carleton: Parr, May 16, 1785. Form for Land Grant Application Renewal, Grimross. May 16, 17, 1785; Rachel Myers to *(probably the local land agent James Peters)* at Grimross Creek, May 30, 1785. RS-108, PANB.

(34) Rachel Myers to Thomas Carleton, Petition, Gagetown, June 15, 1786. RS-108, PANB

(35) Benjamin Myers to Thomas Carleton, Petition, Grimross plot (sic), July 24, 1786, RS-108. PANB.

(36) Rachel Myers to Thomas Carleton. Petition, August 29, 1786 Gage town. RS-108, PANB. Samuel Dickinson, a native of Dutchess County, NY and a "troops guard" in the Loyalist Provincial Corps, settled in Queens County, NB, where he served as a member of the House of Assembly in 1786 and as a Magistrate in 1792. Sharon Dubeau, *New Brunswick Loyalists: A Bicentennial Tribute* (Agincourt, Ontario: Generation Press, 1983), pp. 43-44.

(37) Nick and Helma Mika, *United Empire Loyalists: Pioneers of Upper Canada* (Belleville, Ontario: Mika Publishing Company, 1976), pp. 207-208; Wallace Brown and Hereward Senior, *Victorious in Defeat: The Loyalists in Canada* (Toronto & New York: Methuen Publications, 1984), p. 85.

(38) Neil Mackinnon, *This Unfriendly Soil: The Loyalists Experience in Nova Scotia 1783-1791* (Kingston and Montreal: McGill-Queen's University Press, 1986). pp.158-164; Oscar Zeichner, "The Loyalist Problem in New York After the Revolution", *New York History* (New York Historical Society, XXI, 1940), pp. 296-300.

(39) Rachel Myers to Samuel Dickenson (sic)....#1....1st Survey on Grimross Neck 1 May 1787 *Queens County Land Registers to 1866.* L.R.I.S. Office, Saint John, New Brunswick, Canada. Courtesy of Jan Dexter.

(40) *PAJHS,* XXVII, 43, 253-54; De Sola Pool, *Portraits,* pp. 101-102.

(41) David and Tamar De Sola Pool, *An Old Faith in the New World: Portrait of Shearith Israel 1654-1954* (New York: Columbia University Press, 1955), p. 287, p. 522, n.75.

(42) *Reminiscenses,* pp. 10-11.

(43) Herbert T. Ezekiel and Gaston Lichtenstein, *The History of the Jews of Richmond From 1769 to 1917* (Richmond, Virginia, 1917), et passim. Hereafter cited as Ezekiel and Lichtenstein; Leon Huhner, "The Jews of Virginia From The Earliest Times To the Close of the Eighteenth Century", *PAJHS* (1911), XX, 85-105; "Richmond", *EJ,* XIV, 160-162; Jacob Ezekiel, "The Jews of Richmond", *Studies in American Jewish History,* No. 4, 1894.

(44) Ezekiel and Lichtenstein, p. 240.

(45) Schappes, p. 591, n.2

(46) *PAJHS,* XXVII, 50-51.

(47) Marcus, *U.S. Jewry,* vol. I, 572-573.

(48) Robert W. Reid, *Washington Lodge No. 21, F. & A.M. and Some of Its Members* (New York: Washington Lodge, 1911), p. ix, p. 163. Hereafter cited as Reid; *Proceedings of the Grand Chapter Royal Arch Masons of New York State at Its One Hundred and Seventeenth Annual convention Held Albany* (New York: the Grand chapter, 1914), p. 111. Hereafter cited as *Proceedings.*

(49) *New York City Directory* (1797), p. 256.

(50) Myers joined the Society on May 10, 1797 and was formally admitted and initiated on June 30, 1800. *New York City, Society of Tammany or Columbian Order, Constitution and Roll of Members, 1789-1816,* p. 123; *Minutes of Tammany Society or Columbian Order, 1791-1795,* p. 52. New York Historical Society.

(51) *New York City Directory* (1800), p. 287.

(52) *PAJHS,* XXI, p. 168, p. 89.

(53) *Manumission Society, New York City, Minutes,* Vol. IX, December 7, 1802 and January 18, 1803. MS, New York Historical Society.

(54) Hugh Hastings, ed., *Military Minutes of the Council of Appointment of the State of New York, 1783-1821* (Albany: James B. Lyon, 1901), Vol. II, p. 1089. Hereafter cited as Hastings, *Military Minutes; Proceedings,* p. 110.

(55.) Myers was made a Senior Captain on May 24, 1809. Hastings, *Military Minutes,* Vol. 11, p. 1089; *Proceedings,* p. 110.

(56) *Reminiscenses,* p. 12;   Donald E. Graves, *Field of Glory: The Battle of Crysler's Farm, 1813* (Toronto, ON: Robin Brass Studio, 1999). p. 12. Hereafter cited as Graves, *FOG.*

(57) Joseph L. Blau and Salo W. Baron, eds., *The Jews of the United States 1790-1840: A Documentary History* (New York and London: Columbia University Press, 1963), Vol. II, p. 313. Hereafter cited as Blau and Baron.

(58) Charles A. Brockerway, *One Hundred Years of Aurora Grata, 1808-1908* (Brooklyn, N.Y.: Aurora Grata Consistory, 1908), pp. 2-21; Samuel Harrison Baynard, *History of the Supreme Council, 33: Ancient Accepted Scottish Rite of Freemasonry Northern Masonic Jurisdiction of the United States of America and Its Antecedents* (Boston, MA: Grit Publishing Company, 1938), Vol I, p. 162; Reid, p. 163.

(59) See the *New York City Directories,* 1801-1810, inclusive.

(60) *New York Mayors Court, Calendar of Cases.* Reel IV, 1303; VI, pp. 228-231, pp. 454-475, p. 502, p. 548; VII, pp. 69-70, p. 513, pp. 587-588, p. 718; VIII, pp. 158-159, pp. 220-235; IX, pp. 399-402. Hereafter cited as NYMCCC. Courtesy of the American Jewish Historical Society (AJHS).

(61) NYMCCC. Reel VIII, pp. 98-100. AJHS

(62) *New York City Directory* (1809), p. 280.

(63) NYMCCC *Insolvency Assignments – Insolvent Debtors.* Reel III, pp. 440-462. AJHS

(64) *PAJHS,* XXVII, pp. 394-395.

(65) Hastings, *Military Minutes,* vol. II, p. 1383.

(66) Francis B. Heitman, *Historical Register and Dictionary of the United States Army From Its Organization, September 29, 1789 To March 2, 1903.* Washington: Government Printing Office, 1903. Vol. I, p. 486. Hereafter cited as *HRDUSA.*

(67) *Reminscenses,* p. 13.

(68) Mordecai Myers to Col. P.P. Schuyler, June 7, 1812. *Mordecai Myers Collection,* Clements Library, University of Michigan, Ann Arbor, Michigan. Hereafter cited as *MMC.*

(69) *Plattsburgh Republican,* July 3, 1812. p. 3, col. 4; July 24, 1812, p. 3, col. 2.

(70) Isaac Clark to Mordecai Myers, September 2, 1812. *MMC.*

(71) *Reminiscenses,* pp. 19-26; Charles Boerstler to Myers, February 23, 1813. *MMC.*

(72) Myers to Naphtali Phillips, March 31, 1813. *PAJHS,* XXVII, pp. 396-397. Myers is here paraphrasing from Plutarch's biography of Julius Caesar. He may have read it on his own or as a campaign worker for Aaron Burr heard the reference in discussion. Plutarch was a favorite of Burr's and he was known to recommend the work to family and associates. For the original version, see Plutarch, *The Lives of the Noble Grecians and Romans.* Trans. John Dryden (New York: The Modern Library, 1940), p. 861.

(73) *Reminscenses,* pp. 27-31, pp. 54-55.

(74. *Reminscenses,* pp. 33-37.

(75) *Reminscenses,* pp. 37-39; Graves, *FOG,* pp. 61-62, p. 78, p. 126; The first-hand account of Myers' role in the rescue is in Benson J. Lossing, *Pictorial Field-Book of the War of 1812* (New York: Harper & Brothers, 1869), p. 646, n. 1.

(76) *Reminscenses,* pp. 40-41; See Graves, *FOG* for the definitive account of the battle.

(77) *Reminscenses,* 42-43; Neil B. Yetwin, "Dr. Albon Man: 'A Physician of Large Practice' ", *Franklin Historical Review* (Malone New York: The Franklin County Historical Society and Museum. Vol. 26, 1989), pp. 22-32; Wilhelmina and Clifford Richards, transcribers and editors, "The Recollections of Susan Man McCulloch", *Old Fort News,* Allen County-Fort Wayne Historical Society, Fort Wayne, Indiana. Vol. 44, No. 3, 1981, p. 8, hereafter cited as McCulloch; Cummings, a Georgia native, received a musket ball through the thigh and took months to recover. See Graves, *FOG,* p. 220, p. 401, n. 73; John C. Fredericksen, *The War of 1812 in Person: Fifteen Accounts by U.S. Regulars, Volunteers and Militiamen* (Jefferson, N.C.: McFarland & Company, Inc. Publishers, 2010), pp. 143-145; Heitman, *HRDUSA,* vol. I, p. 344; Francis B. Heitman, *Historical Register of the United States Army, From Its Organization, September 29, 1789 to September 29, 1889* (Washington, D.C., 1890), p. 212, cited hereafter as *HRUSA;* Preston, though permanently disabled from his injuries, went on to serve as governor of

Virginia, 1816-1819. Robert Alonzo Brock and Virgil Anson Lewis, *Virginia and Virginians: Eminent Virginians* (Richmond and Toledo: H.H. Hardesy, Publishers, 1888), Vol. I, pp. 131-132; Heitman, *HDRUSA,* vol. I, p. 806.

(78) Myers' injury was a proximal humerus fracture. Destroyed tendons affects shoulder motion, as the humerus connects the upper arm bone connecting the shoulder to the elbow. Mordecai Myers, Reg. of Enl. – U.S.A. - Record Group No. 94, Military File. Vol. 16, p. 161, #5353; Mordecai Myers, Capt. U.S. I. File No. 27, 240, Vol. 2, p. 293. National Archives, Washington, D.C., cited hereafter as *Military File;* also see Catalina Mason Myers Julian James, *Biographical Sketches of the Bailey-Myers-Mason Families, 1776-1905* (Washington, D.C., 1908), p. 19, hereafter cited as *Biographical Sketches.* The late Walter Mason, whose grandfather purchased the property from Albon Man's grandson, took the editor on a tour of the house and grounds, indicating the room where Myers was treated. The window that Charlotte Bailey covered with a curtain is still in its original location.

(79) McCulloch, p. 8; *Plattsburgh Republican,* January 29, 1814. p. 3,col. 4.

(80) *Military File,* Book 405.

(81) *Reminiscenses,* p. 45.

(82) *Reminiscenses,* p. 44; Graves, FOG, p. 328; Military File, Book 405.

(83) James Kent to Mordecai Myers, October 20, 1814. *Theodorus Bailey Myers Collection,* #2132. New York Public Library. Hereafter cited as *TBMC.*

(84) C.K. Gardner to Mordecai Myers, November 16, 1814. *MMC;* Hugh Hastings, ed., *Public Papers of Daniel D. Tompkins, Governor of New York 1807-1817: Military* (New York and Albany: Wynkoop Hallenbeck Crawford, 1898), Vol. I, p. 328. Hereafter cited as Hastings, *Tompkins Papers.*

(85) Hugh R. Martin to Mordecai Myers, December 16, 1814. *MMC.*

(86) Major Richard Malcolm to Mordecai Myers, January 22, 1815. *MMC.*

(87) *Military File,* Book 405.

(88) Lt. James A. Chambers to Mordecai Myers, March 1, 1815. *MMC.*

(89) Robert S. Quimby, *The U.S. Army in the War of 1812: An Operational and Command Study.* East Lansing, Michigan: Michigan State University Press, 1997. Vol. I, pp. 77-78.

(90) Willoughby Morgan to Mordecai Myers, March 21, 1815. *MMC.*

(91) Heitman, *HRDUSA*, Vol. I, p. 486; M. Myers, *Military Pension File:* "Invalid". SC-27240: War of 1812. File No. 27, p. 240, Vol. 2, p. 293. National Archives, Washington, D.C. Myers's offical date of record for retirement was listed as June 15, 1815, however, he was retained in active service and credited with pay until September 10, 1815. Myers's pension commenced, retroactive, to June 16, 1815.

(92) *Reminiscenses*, p. 45.

(93) Harry L. Coles, *The War of 1812* (Chicago and London: The University of Chicago Press, 1965), pp. 238-242. Hereafter cited as Coles.

(94) Coles, pp. 268-269.

(95) Donald R. Hickey, *The War of 1812: A Forgotten Conflict* (Urbana and Chicago: University of Illinois Press, 1989), p. 303.

(96) *PAJHS*, XXVII, p. 398.

(97) *Reminiscenses*, p. 46; Heitman, *HRUSA*, p. 677; Alina Marie Lindegren. *A History of the Land Bonus of the War of 1812* (University of Wisconsin, Unpublished Masters Thesis, 1922), p. 17.

(98) Ira Cohen, "The Auction system in the Port of New York, 1817-1837." *The Business History Review.* Vol. 45, No. 4 (Winter, 1971), pp. 38-40. Hereafter cited as Cohen.

(99) Judge William Bailey to Mordecai Myers, November 8, 1821. *MMC.* Vol. II, p. 93.

(100) Theodorus Bailey to Judge William Bailey, April 3, 1823. *MMC,* Vol. II, p.79. Charlotte's brother Theodorus Bailey (1805-1877) was at this time a recent enlistee in the U.S. Navy. He was serving then on a receiving ship used in New York Harbor to house newly recruited sailors before they were assigned to a crew. Bailey served with distinction in the Mexican War and the Civil War, attaining the rank of Rear Admiral.

(101) Reid, 52-53; *Proceedings*, 111.

(102) *Record in the Case of the United States of America Versus Fernando M. Arrendondo and Others. Supreme Court of the United States.* January Term, 1831 (Washington: Printed By duff Green, 1831), pp. 129-132; Maurice G. Baxter, *Daniel Webster & the Supreme Court* (Amherst, MA: University of Massachusetts Press, 1966), pp. 143-145.

(103) "M. Myers & Co., 107 Water St.", *New York Enquirer,* July 7, 1826, p. 2. Quoted in Blau and Baron, vol. I, p. 284, #211; See also Marcus, *U.S. Jewry,* Vol. I, pp. 168-169.

(104) *The Military Society of the War of 1812 – Annals, Regulations, and Roster.* Secretary and Adjutant's Office, March 12, 1895. pp. 1-4.

(105) Harlan Whately, "The History of the VCASNY (Veteran Corps of Artillery of the State of New York) and the Military Society of the War of 1812", Part I, February 19, 2008, pp.1-5. On Sunday, May 21, 2006 the Color Guard of the VCA participated in Congregation Shearith Israel's annual memorial service at the synagogue's Chatham Square Cemetery. They honored Mordecai Myers memory by decorating the grave of his mother, Rachel Myers, with the American flag. See "Memorial Service for American-Jewish War Heroes 2006", *Veteran Corps of Artillery, State of New York,* p. 1 (retrieved from www.jag – consulting.com/VCA/PhotoGallery.php?album=3).

(106) McCulloch, p. 8.

(107) See William Preston Vaughn, *The Antimasonic Party in the United States, 1826-1843* (Lexington, Kentucky: University Press of Kentucky, 1983), et passim; Michael F. Holt, "The Antimasonic and Know Nothing Parties", in Arthur M. Schlesinger, ed., *History of U.S. Political Parties* (New York: Chelsea House Publishers, 1973), Vol. I, pp. 575-620.

(108) Reid, pp. 55-58, p. 65.

(109) Cohen, pp. 505-510.

(110) Robert V. Remini, "The Albany Regency", New York History, 39 (Oct. 1958), 341-355; Also see Arthur M. Schlesinger, Jr., *The Age of Jackson* (New York: Little, Brown & Company, 1945), et passim.

(111) *Morning Courier,* November 6, 1828, p. 2.

(112) *Journal of the Assembly of the State of New-York at their Fifty-First Session and Held at the Capitol in the City of Albany, 1828* (Albany: Printed by E. Croswell, Printer to the State, 1829), Vol. I, pp. 128-129; Vol. II, pp. 688-689, pp. 700-701, p. 854. Hereafter cited as *JASNY.*

113. *JASNY* (1829). Vol. I, pp. 76-77, p. 95, p. 106, pp. 121-122, p. 161, pp. 202-203, p. 303, p. 381, pp. 391-394, pp. 479-478, pp. 551-552, p. 647, p. 655, p. 718, pp. 752-757, p. 784, pp. 794-795; Vol. II, pp. 845-849, p. 857, p. 906, p. 909, p. 917, p. 921, p. 935, p. 947, p. 1004, p. 1135.

(114) George Bliss, Peter Olney, & William Whitney (compilers), *Special and Local Laws Affecting Public Interests in the City of New York, In Force on January 1, 1880* (Albany: Charles Van Benthuysen & Sons, 1880), Vol. II, pp. 2006-2008.

(115) *JASNY* (1831), p. 50.

(116) *Biographical Sketches,* p. 13; *JASNY* (1832), vol. I, 22, 61, 65, 140-141, 144, 161, 418, 429; *Documents of the Assembly of the State of New York, Fifty-Fifth Session, 1832* (Albany: E. Croswell, Printer to the State, 1832), Vol. III, From No. 175-275 Inclusive, pp. 1-2; No. 222 (March 17, 1832); No. 44, pp. 1-3, 7. Hereafter cited as *DASNY;* Generous land bounties were offered to veterans of the Mexican War under the Ten Regiments Act of 1847. This angered surviving veterans of the War of 1812 who formed lobbying groups to rectify what they saw as a gross injustice. But there had been demands for bounty lands since 1826 when Myers' "Committee of the Officers of the War of 1812" sent Congress their petition. Over the next 30 years similar petitions came from Indiana, Maine, Kentucky, Mississippi, Arkansas, Virginia, Maryland, Ohio, New Hampshire, Rhode Island, Jew Jersey and Wisconsin. New York alone sent 47 petitions with 7100 signatures, and a "National Convention of the Soldiers of the War of 1812" was held at Philadelphia, "the chief center for bounty land agitation." James W. Oberly, "Gray-Haired Lobbyists: War of 1812 Veterans and the Politics of Bounty Land Grants", *Journal of the Early Republic,* Vol. 5, No.1 (Spring, 1985), pp. 38-41.

(117) James S. Kabala, " 'Theocrats' vs. 'Infidels': Marginalized Worldviews and Legislative Prayer in 1830s New York." *Journal of Church and State,* Volume 51, No. 1 (2009), pp. 92-93. Hereafter cited as Kabala.

(118) The "Moulton-Myers Report" has been reprinted in its entirety in Joseph L. Blau, ed., *Cornerstones of Religious Freedom in America* (Boston: The Beacon Press, 1950), pp. 136-156.

(119) James R. Willson, *Prince Messiah's Claims To Dominion Over All Governments: And the Disregard of His Authority By the United States, In the Federal Constitution* (Albany, N.Y.: Packard, Hoffman and White, 1832), pp. 11-12, pp. 17-30, pp. 34-35. Hereafter cited as Willson.

(120) *Proceedings,* p. 111.

(121) *Willson,* p. 36.

(122) Kabala, 78-79; *Albany Evening Journal,* January 3, 1832, n.p.

(123) *JASNY* (1833), 97.

(124) *JASNY* (1834), 67-68, 881-883; *DASNY,* Vol. IV, 13-14.

(125) Solomon Southwick (pseudonym "Sherlock"), *A Layman's Apology, For The Appointment of Clerical Chaplains By the Legislature of The State of New York In A Series of Letters, Addressed To Thomas Hertell, 1833 – Signed* (Albany, New York: Hoffman & White, 1834), p. xxv. Hereafter cited as Southwick.

(126) Southwick, p. 78.

(127) Southwick, p. 282.

(128) *Reminscenses,* p. 46.

(129) Myers borrowed $5400 for the purchase of the property from the estate of landowner-moneylender William James, the grandfather of novelist Henry James and philosopher William James. Myers to Gideon Hawley (executor of James' estate), October 4, 1836. Mordecai Myers, *Letterbook, 1835-1841.* Hereafter cited as *Letterbook.* Courtesy of the late Dr. John S. Hoes; See also James Edward Leath, *The Story of the Old House: Columbia county's "House of History"* (Kinderhook, N.Y.: the Columbia County Historical Society, 1947), p. 17. Hereafter cited as Leath.

(130) Henry James Genet (1800-1872) was the eldest son of Edward Charles "Citizen" Genet (1763-1834), French ambassador to the U.S. during the French Revolution. The elder Genet had compromised American neutrality by encouraging France's wars with Spain and Great Britain. When it became clear that he would likely be guillotined if he returned to France, he sought asylum in the U.S., married a daughter of New York Governor George Clinton and settled on a farm in East Greenbush, NY called "Prospect Hill", where Henry James Genet was born.

(131) M. Myers to Genet, June 22, 1835. *Letterbook.*

(132) Edward A. Collier, *A History of Old Kinderhook* (New York and London: G.P. Putnam's Sons, 1914), p. 300. Hereafter cited as Collier; Kinderhook *Sentinel,* October 26, 1837, p. 2.

(133) *Biographical Sketches,* pp. 35-36; Leath, pp. 18-19.

(134) M.Myers to Henry Thomas, no date, *Letterbook.* Thomas was one of 124 soldiers who participated in the "Caledonia" and "Detroit" incident on October 9, 1812. See Lossing, p. 386, n. 5.

(135) *Kinderhook Sentinel,* May 9, 1839, p.2.

(136) Myers' address to Van Buren was published in full in the *Kinderhook Sentinel,* July 18, 1839, p.2. An edited version of the address appears in Collier, pp. 259-264.

(137) Leath, p. 18.

(138) *Kinderhook Sentinel,* July 23, 1840, p. 2.

(139) Myers to Mrs. (William) Hatch, July 5, 1840. *Letterbook.*

(140) Myers to Gideon Hawley, April 12, 1840. *Letterbook;* For the Panic of 1837 and its impact, see Samuel Rezneck, "The Social History of an American Depression, 1837-1843", *The American Historical Review.* Vol. XL, No. 4, July, 1935. pp. 662-687.

(141) Myers to Winfield Scott and Isaac Varian, November 16, 1841. *Letterbook;* Silas Wright to Myers, December 5, 1841. *TBMC.*

(142) *Kinderhook Sentinel,* November 7, 1845, p. 2.

(143) *Kinderhook Sentinel,* February 15, 1848, p. 2.

(144) Leath, p. 18.

(145) *Reminscenses,* p. 48.

(146) The site of the Myers' first Schenectady residence as of 2013 is occupied by Habitat for Humanity, Inc., 115 North Broadway, which was called Centre Street until 1854, in what was then the city's 4th Ward. The 1855 *Census of Schenectady County* (p. 160, #191, Schenectady County Historical Society, hereafter cited as SCHS) states that the Myers had been living in the city for 7 years.

(147) The Myers also employed 2 Irish servants, Bridget Doyle and Mary Gesser. See *1850 Census – City of Schenectady,* transcribed by H. Ritchie, p. 124. SCHS.

(148) Hugh R. Martin was commissioned a captain in the *13th Infantry* on March 12, 1812, and was promoted to Major of the *22nd Infantry* on September 12, 1814. He was honorably discharged June 15, 1815. Martin died at Schenectady on May 11, 1848, at age 69. Heitman, *HRUSA,* p. 692; Hastings, *Tompkins Papers,* Vol.II , pp. 411-412; *Schenectady Reflector,* May 19, 1848, p. 2.

(149) John Keyes Paige became a First Lieutenant in the *13th Infantry* on March 17, 1812 and captain on May 13, 1813. He was honorably discharged June 15, 1815. Paige was the first District Attorney of Schenectady County, Clerk of the New York State Supreme Court, Presidential Elector in 1844 and one-time Mayor of Albany, NY. Heitman, *HDRUSA,* Vol. I, 765; *Schenectady Reflector,* December 18, 1857, p.2.

(150) Myers and 4 others were nominated by a 3-man committee of the Democratic-Republican Electors to represent Schenectady's 4th Ward at the County Convention held at Schenectady's Fuller Hotel on the following Saturday, September 14, 1850. *Schenectady Reflector,* September 6, 1850, p. 3.

(151) *Schenectady Reflector,* April 14, 1851, p. 3. The details of Myers' first term as mayor of Schenectady were published in the Common Council Minutes and reprinted in the *Schenectady Reflector,* from April 11, 1851, to April 8, 1852, inclusive, on p. 3 of each issue.

(152) *Diary or Journal of Judge Thomas Palmer, Schenectady, NY, Jan. 1, 1853 to Dec. 31, 1853.* Book IV, p. 431. The Special Collections of the Schaffer Library, Union College, Schenectady, NY. (SCSLUC)

(153) *Proceedings*, p. 111.

(154) Reminiscenses, p. 5. Myers began the letter(s) on February 21, 1853.

(155) *Schenectady Reflector*, March 31 and April 7, 1854, each on p. 3; *Journal of the Common Council of the City of Schenectady From April 6, 1854, To April 4, 1855* (Keyser, Printer: 1855), pp. 3-10. Hereafter cited as *Journal*. (SCSLUC)

(156) *Journal*, pp. 50-57; *Schenectady Evening Star*, January 7,1855, p. 3.

(157) W.T.Bailey, *Richfield Springs and Vicinity. Historical, Biographical, and Descriptive* (New York and Chicago: A.S. Barnes & Company, 1874), p. 67.

(158) Schenectady *Reflector & Democrat*, October 19, 1860, p.3; Interestingly, an identical announcement of Myers' candidacy appeared in *The Jewish Messenger*, October 19, 1860, p. 182. Courtesy of the American Jewish Archives, Cincinnati, Ohio.

(159) Mordecai Myers to Theodorus Bailey Myers, October 26, 1860. *MMC*, Vol.II, p.54.

(160) For Lossing's brief visit to Schenectady, see Benson John Lossing, *Notebooks and Journals*. Microfilm LS 1110-1128. The Huntington Library, San Marino, California; also see Lossing, p.646, n.1; pp. 653-654, n.6.

(161) *Biographical Sketches*, p. 11.

(162) *Biographical Sketches*, pp. 15-16.

(163) *Schenectady Democrat & Reflector*, July 16, 1863, p. 2.

(164) *Schenectady Democrat & Reflector*, October 8, 1863, p. 2. Consumption (TB) was common in Schenectady as in many American cities of the 19th century, and the city had been struck a number of times by cholera, but the cause of death for either of Myers' sons is not known.

(165) Mordecai Myers to Charlotte M. Jackson, May 17, 1864. *MMC*, Vol. II, p.60. Charlotte, named after her mother, was married to architect Thomas R. Jackson (1826-1901), whose work included the designs for Trinity Church in New York, the Brooklyn Theater, and the first *New York Times* building at 41 Park Row.

(166) *Schenectady Democrat & Reflector*, February 7, 1865, p.2.

(167) *Schenectady City Directories*, 1866-68; For material on Edgar M. Jenkins see his obituary, *Schenectady Union Star*, December 31, 1914, p.1.

(168) Myers purchased the property, now 231 Union Street in Schenectady Historic Stockade District, on August 16, 1869, for $5150. It is still a private residence. *Deeds*. Book 52, p. 362. Office of the Schenectady County Clerk.

(169) "Relating to the Military history of our Grandfather Major Myers", unsigned, undated typed manuscript, MMC (Vol. II, pp. 71-72).

(170) Mordecai Myers to Theodorus Bailey Myers, December 31, 1869. *MMC*, Vol. II, p.96.

(171) *Schenectady Evening Star*, January 20, 21, 25, 1871, each on p. 3; *Schenectady Daily Union*, January 24-25, 1871, each on p.3; excerpts from these extensive obituaries appeared in the *Albany Argus* (one-time organ of Martin Van Buren's "Albany Regency"), January 20, 1871, and in the *New York Times*, January 25, 1871.

(172) Allen B. Mann, *A History of Congregation Gates of Heaven*. Undated typed manuscript, circa 1938-1940. Courtesy Congregation Gates of Heaven.

(173) Schenectady Surrogate Court Records Room, Box No. 87.

(174) *Schenectady Evening Star*, January 25, 1871, p.3.

Part Two:
The Reminiscences of Mordecai Myers, 1780 – 1814, as published in 1900;
re-keyed,
with notes by Neil B. Yetwin
in 2013.

*The location of the original pagebreaks
and page numbers of the text of
the 1900 edition
are indicated in*
[bold brackets]
for reference comparison;
additionally,
*several clarifying editorial notes (2013) appear
within the main text, also in brackets.*

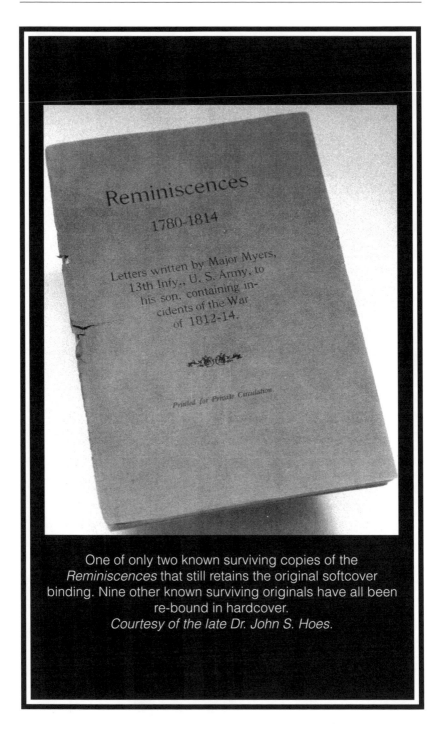

One of only two known surviving copies of the
Reminiscences that still retains the original softcover
binding. Nine other known surviving originals have all been
re-bound in hardcover.
Courtesy of the late Dr. John S. Hoes.

[Front Cover]

Reminiscences
1780-1814
Letters written by Major Myers,
13th Infy., U. S. Army,
to his son,
containing incidents of the War of 1812-14.

Printed for Private Circulation.

—— —— —— —— —— ——

[a front end sheet page; not numbered]

One hundred copies of this pamphlet have been printed.

No. _____

Presented to [name to be inscribed]

—— —— —— —— —— ——

[Title page; not numbered, but Page 1]

Reminiscences
1780 to 1814
INCLUDING INCIDENTS IN THE WAR
OF 1812-14

LETTERS PERTAINING TO HIS EARLY LIFE

Written by MAJOR MYERS, 13th Infantry, U.S. Army,
to
HIS SON

————————————

THE CRANE COMPANY
1411 G ST.
WA SHINGTON, D.C.
1900

Page 4, not numbred, of the original *Reminiscences*
showing the portrait of Myers by John Wesley Jarvis, and
Myers' distinctive signature below. *Courtesy of the
Schenectady County Historical Society (SCHS).*

[Page 5]

EXTRACTS FROM LETTERS WRITTEN BY
MAJOR MYERS,
PERTAINING TO HIS EARLY LIFE.

~~~~~~~~~~~~~~~

SCHENECTADY,
February 21, 1853

MY DEAR SON: — [1]

In accordance with your expressed wish, I will give you a condensed outline of the events of my long and varied life, beginning with a short account of family affairs. My father was a Hungarian, and my mother an Austrian, by birth.[2] They sailed from Helvoetsluys, in Holland,[3] and arrived in New York in the year 1750. They soon after removed to Newport,[4] where I was born on the 31st of May, 1776[5] — two months before the signing of The Declaration of Independence. My father spoke and wrote all the living languages,[6] and, at Newport, he became the friend of the Rev. Dr. Styles,[7] afterwards President of Yale College. In November, 1776,[8] before having reached his fortieth year, my father died, beloved and respected by all who knew him.

The British evacuated Newport in 1779, and, in 1780, my mother decided to return to New York.[9] The winter of this year was known long after as the "hard winter." The snow was deeper than it had ever been known to be before on the Continent; and I have seen nothing like it since. The Bay of New York was so firmly frozen over that the British marched troops and removed heavy guns on the ice to Staten Island. We lived quietly in New York until 1783, when Great Britain was compelled, by her own necessities, and by the persevering bravery of the Americans, to acknowledge the American Colonies to be free and independent States. Preparation were being made to withdraw the forces, and the British, in common justice to those Americans who, unfortunately for themselves, had embraced the British cause, offered homes in Nova Scotia to all who would sign Articles of Association. My mother agreed to this, and early in May we left New York in a British ship for our new home. We spent several happy years in Nova Scotia, where a pleasant society was formed.[10] Our nearest neighbors were Captain Richard Lippincott and his wife, a Quaker. He had been a Captain in the British Army, and. as such, he was ordered, sometime in the year 1782, to exchange the American [Page 6] Captain Huddy, then a prisoner in New York, for a British officer of equal rank, who was to be delivered to him on the Jersey shore at, or near, Bergen Point; but on the way, he landed at Gibbet Island, and ordered his prisoner to prepare for death.[11]

Huddy considered it a jest, and endeavored to laugh it off; but Lippincort directed his Negro servant to prepare a halter, and Huddy was actually hung to the limb of a tree. Lippincott then returned to New York, of course, without the British officer whom he should have brought in the place of Huddy.

Great excitement was felt in New York, as well as in the American camp. Lippincott was arrested and imprisoned, and General Washington demanded his delivery to the Americans, but was answered that the trial would take place under British law. All expected the condemnation and execution of the criminal, but at about this time, the Treaty of Peace was signed and ratified, whereby our independence was acknowledged, and Lippincott was either acquitted or pardoned, when he went to Nova Scotia. There was a story current at the time about the notorious Benedict Arnold. He has become a merchant in St. Johns, and, in connection with Monson Hoyt, had imported from England, not knowing what would be required in the new country.[12] Consequently, he accumulated a large quantity of unsalable goods, which were, however, highly insured in England. It is said that he hired a woman, for a new dress, to set fire to the building in which his goods were stored. But the plot was discovered at the trial for the recovery of the insurance, and Arnold was obliged to leave the country.

In 1787 we returned to New York. On the voyage we stopped in New haven, where we received an additional passenger — Captain John Paul Jones.[13] I remember him as a man of medium height, rather stout and well set, with dark hair and eyes. In conversation, free and easy; and in manner, rather bombastic. He seemed to me to look more like a German than a Scotchman. At New Haven I saw a young man tied to a post and publicly whipped for horse-stealing.

I had seen, when in New York, Hessian Soldiers run the gauntlet; and a man under the "mild and humane" British laws, standing in the pillory, cropped and branded for stealing a loaf of bread from a baker's widow.

In New York we took a house owned by the excellent Mr. [Page 7] Randall. He had begun life a poor sailor boy; but, by his own exertions, he became a very rich merchant. He had been aided by a sea-faring man, and he amply returned the favors by leaving the bulk of his fortune to be used for the founding of an institution for the benefit of disabled and superannuated sailors. That fund forms the basis of the Sailors' Snug Harbor on Staten Island.(14)

At this time, in the year 1787, New York contained 33,000 inhabitants. The city was still in rather a dilapidated state, not having, as yet, recovered from the effects of the war. The country was governed under the old Articles of Confederation formed during the Revolutionary War merely to establish a union of force and action without defining or limiting the rights or powers of the general government of the States. Our commerce, both foreign and domestic, was very small and much embarrassed. A vessel and cargo going from one State to another was compelled to clear, enter, and secure duties.

Each State had its proper currency which would not pass in the adjoining States; every kind of property was low and money scarce.

Lots of ground on Broadway between Grand and Great James Streets were sold at from twenty to twenty-five pounds; but corner lots were held at a little more than two pounds higher; in all the cross streets the price of lots was from ten to twelve pounds each. Flour was four dollars a barrel; beef, four cents per pound; butter, from eight to ten cents; a cart load of hickory wood, seven shillings, cartage or sawing, one shilling per load; and everything in proportion. What is now the Park was then an unfenced space, so muddy in wet weather that, to cross it, one must go ankle deep in mud. Where the City Hall now stands, was a range of wooden buildings one story high, with a common picket fence in front; this was the City and County Alms House. There was an unfinished stone building in range of the above on Broadway called the Bridewell; it was afterwards finished and used as a prison. On the left was the old Provost Prison of the Revolution, where many American prisoners were either hung in the cells by Cunningham, the cruel keeper; or were suffered to starve.[15] After the peace, it became a debtors prison; then, it was newly modeled, as it now stands, the Hall of Records. There was scarcely a house in view west and north of that location, excepting the New York Hospital then in a dilapidated state, on Broadway. The new Constitution [Page 8] of the United States had been agreed to in convention, but had not, as yet, been adopted by all the States. The French revolution had burst forth, and party spirit ran high. Two great political parties were forming. Commodore Francis Nicholson[16] was the first President of the Democratic Society, and Thomas Jefferson[17] was its great leader. Alexander Hamilton[18] lead the Federal party. Jefferson and the Democrats were jealous of the power of the President by the new Constitution; and they considered the office nearly equal to that of king. They feared the establishment of a concentrated. luxurious, and extravagant government; a great controlling political institution; and a union of Church and State. They believed that General Hamilton, Timothy Pickering,[19] Oliver Wolcott[20] and others were trying to introduce a limited monarchy in disguise. Among the propositions made to the Convention, General Hamilton suggested that senators should be elected for life, and that the President should choose his successor. Mobs and riots were common, the Democratic printing office of Greenleaf was attacked,[21] and the type scattered in the street; but after great and long excitement, the Constitution was ratified by most of the States. The convention of this State (N. Y.), having peace and the public interest in view, met at Newbury,[22] and, after long debate, ratified the Constitution. At the same time, a declaration of Reserved State Rights was made, a copy of which may be found in the Assembly documents of 1833, placed there at my request, I having found the original in the office of the Secretary of State, after long search. George Washington was elected President of the United States, and all the States elected Congressmen, who now assembled at the old City Hall in New York — it stood at Wall Street, where the Custom House now stands. I recollect seeing Chancellor Livingstone[22] administer the oath of office to General Washington on a Bible which is still in a state of good preservation, and in the possession of St. John's Lodge, No. 1,. New York, where it is held as a relic of times past. Although the Constitution was adopted, adopted, and civil government in full force in America, and the revenue and property rising in value, the two great political parties existed.

The Federalists went so far as to invent a black cockade as a distinctive badge, and the Democrats,. or Anti-Federalists, not adopting it, were often insulted, and even pushed off the pavement.

The greatest exertions were made by each party to elect its [Page 9] members for the Legislature and Congress, and with varying success.

This continued until 1799, when the Democrats gained the ascendency; and, within a few years thereafter, we find thirteen amendments to the Constitution as originally adopted, all of which are in favor of civil liberty. The State Convention having ratified the Constitution of the United States accompanied by all instrument signed by all its members, and known as the Reserved Rights of the State of New York, both parties united in getting up a pageant such as had never been seen in New York. The corporation invited all the mechanical, civil, and charitable societies to appoint committees to make arrangements for a general procession. Each was allowed to make its own preparations, which were done on an extensive and brilliant plan. Early on the day appointed, each society assembled at the Battery.

The Governor, both branches of the Legislature, the judges, the members of the bar, the officers of the courts, officers of the army and navy, the Mayor and corporation, and other city officers, took part in the procession. All the mechanical trades were represented by men engaged at their respective occupations in

cars mounted on trucks. The ship carpenters contributed a miniature frigate of thirty-six guns, completely rigged, armed, and
manned. She was called the "United States," and commanded by
Commodore Francis Nicholson. Her crew consisted of young lads
dressed as sailors, her sails were loosed and sheeted home, and
her guns were loaded and fired. This attracted great attention.
During the day, the streets were crowded with people, and the
houses were decorated with flags.[24]

In the evening, Colonel Sebastian Bowman,[25] a Prussian officer of the Revolution, and soon after appointed first postmaster
of New York by General Washington, prepared extensive fire-
works at the old fort, Bowling Green, and at other places. This
was, I believe, the first exhibition of fire-works in New York. All
places of amusement were opened and thronged. Most of the
houses and public buildings were illuminated. It was near morning before the streets were again quiet. Thus was consummated
the foundation of our excellent Civil Government, and the great
Republic of thirteen free and independent United States.

Distraction to the brain that would conceive the idea of a [**Page
10**] separation of the Union, and palsied the hand that would
break one link of the Heaven-wrought chain![26]

The following years of my life were spent in New York, and in
Richmond, Virginia.[27] During my residence in Richmond, I
made many valuable acquaintances, among whom was Judge
Wythe,[28] one of the signers of the Declaration of Independence.
He, and all his family, were poisoned by one of the household
servants, a black boy, who had been treated rather as a pet than
otherwise. The younger members of the family recovered, but
the old gentleman died. The boy was hung; he prayed, and sang
psalms under the gallows, and said that the had always been well
treated by the family, but that the Devil had prompted him to do
the wicked deed; and (as frequently happens in such cases) his
Savior had appeared to him the night before, and pronounced his
pardon.

I became an active politician just before the election of 1793,[29]
which put a period to the despotic reign of John Adams (the
elder) and placed Thomas Jefferson in the Presidential Chair in

the following year. The Federal party was composed in part of the old Tories of the Revolution, and the rich merchants, and traders, who boasted of having all the wealth, talent, and respectability of the American people. Congress, under the Federal Administration, had passed several oppressive laws, abridging the public rights. The Alien and Sedition Laws[30] were particularly objected to by the Democratic party. The President had power to transport or imprison, without trial, any suspected person, and he actually did imprison many valuable citizens, for speaking disrespectfully of him as an individual. Several of these were editors.

The warm partisans of the dominant party would neither deal with or employ those who differed from them in politics. All the offices from highest to lowest were held exclusively by the Federal party. Many of the merchants stooped so low as to discharge clerks, cartmen, and others who differed with them in politics. The measure of wrongs was filled to overflowing. The parties, numerically, were nearly equal; but the wealth and patronage were against the Democrats.

One evening three gentlemen met at the house of Brockholst Livingstone,[31] in Broadway — Mr. Livingstone, General Morgan Lewis,[32] and Aaron Burr.[33] The wrongs of the people was the subject of conversation. Mr. Burr said, "We must, at the next election, put a period to this 'reign of terror'". The others agreed that [Page 11] this was desirable, but saw no way to bring it about. Mr. Burr said, "We must carry the City, and that will give us the majority in the Legislature; and the State of New York being Democratic, will carry the Union, and transfer to the Democrats all the power of patronage of the government."

The other gentlemen thought this a brilliant plan, but did not see how it would be possible to gain the ascendency in the City. Mr. Burr took pen and paper and made out an Assembly ticket, heading it with the names of Gov. George Clinton,[34] Gen. Horatio Gates,[35] Col. Willett,[36] Henry Rutgers,[37] Brockholst Livingstone, Ezekiel Robbins,[38] Aaron Burr, etc., making the whole ticket eleven members. Mr. Livingstone observed that many of these gentlemen would not agree to serve, and that, if they should, it would not be easy to get them nominated and elected.

Mr. Burr requested the gentlemen to discuss the question in a week from that night; but he said, "Mr. Livingston, you and I call agree at once: I will agree to serve my Country on this occasion, and I am sure that you will not refuse." He answered, "No; if the rest will serve." The party separated, feeling great ardor in the cause.

In the course of the week, Mr. Burr called on all the other gentlemen, and with his usual eloquence, and argumentative powers, induced them all to serve. At the end of the week, the three gentlemen met according to agreement, and Mr. Burr reported the assent of all. He next proposed to call a general meeting at Tammany Hall,[39] and said, "As soon as the room begins to fill up, I will nominate Daniel Smith[40] as chairman, and put the question quickly. Daniel being in the chair, you must each nominate one member, I will nominate one, and Fairley,[41] Miller,[42] Van Wyck,[43] and others will nominate, and Daniel must put the question quickly on the names, and, in this way, we will get them nominated. We must then have some-inspiring speeches, close the meeting, and retire. We must then have a caucus and invite some of our most active and patriotic Democrats, both young and old, appoint meetings in the different wards, select speakers to address each, and keep up frequent meetings at Tammany Hall until the election. We will put down the monster Federalism, and bring the country back to pure Democratic principles." The whole plan succeeded, and the civil revolution was brought about.

I give you an account of what took place at Mr. Livingstone's [Page 12] as it was related to me by Gen. Morgan Lewis, and of the after proceedings on my own authority — I being one of the actors. I accompanied Aaron Burr to several meetings which he addressed. I was one of those selected to address the people at Tammany Hall and in the wards. The general election was carried on with great energy by both parties.

Our organization was completed by dividing the city into small districts with a committee appointed to each, whose duty it was to canvas its district and ascertain the political opinion of each voter by going from house to house, seeing and conversing with as many as possible, and enquiring the politics of such as we could not see. The district committees reported their strength at the ward meetings, the names were called off with marginal notes stating whether good, bad, or doubtful. so that at the general meeting, we could determine very nearly what would be the result in the city; and the result of the election destroyed the hydra monster, Federalism.

After about sixty ballots taken in the House of Representatives, Thomas Jefferson and Aaron Burr had an equal vote. At that time, it was not customary to designate which should be President, and which Vice-President. The candidate having the greatest number of votes was chosen President, and the one having the next to the greatest vote, Vice-President. I think it was on the sixty-first ballot that Caesar Rodney,[44] a Representative from Delaware, left his seat, and this gave Thomas Jefferson a majority of one vote. Thus, Aaron Burr was elected Vice-President. At about this time, I joined Captain John Swarthout's[45] artillery company, in which I served for six years. I was then appointed Lieutenant of Infantry in Captain James Cheatham's[46] Company, and was, at the next meeting of the Council, promoted to the rank of Captain in Colonel Van Buren's[47] regiment. I became senior Captain of the regiment, and had command of a battalion; but before-the-commissions were made out for the promotions, (Major, in my case) I left the regiment. At the particular request of D. D. Tornpkins,[48] then Governor, I became a student of Military Tactics, etc.,. under Colonel de la Croix,[49] and remained under his instruction for about two years, until I was appointed Captain in the United States Infantry. I was then a member of a club of forty gentlemen[50] including Governor Tompkins, Lieut.-Gov. Broome,[51] Major Fairley, Dewitt Clinton,[52] Sylvanus Miller, Pierre C. [Page 13] Van Wyck, Henry Remsen,[53] James Cheatham, Daniel Smith, Henry Rutgers, and others.

In 1812, the storm of the war was gathering and I decided to ask for a commission as I was prepared by study for the military profession. I applied for a commission in the army, referring to the Vice-President and several Congressmen from this State; and, in a few days, I received a Captain's Commission, and was assigned to the 13th Regiment of U. S. Infantry, commanded by Colonel Peter P. Schuyler.[54] I now entered the army with rank far beneath what I might have had, as my friend, Governor Tompkins, told me, if I had asked the assistance of my friends. I said that I thought my rank as high as it should be on entering, and that I should prefer to gain my promotion by service. He told me that I would find very many above me who would be more fit to obey than to command; and so I found it, for great ignorance of military tactics prevailed in the new army, and the old one was not far beyond us in field duty. However, things soon took a better aspect, for we had some educated and scientific officers, and all soon improved who were capable of improvement.

I approach, with diffidence, a sketch of my military career, as I cannot extend it so as to give a general account of the war; but must confine myself to the movements in which I myself took part.

At an interview with Colonel Schuyler, he said, "The State of New York comprises five recruiting districts., four of these will be commanded by the field officers of the regiment, and I appoint you to the command of the fifth." He then asked me if I knew of any young officer of talent who would make a good Adjutant. I was pleased with the opportunity to recommend my young friend, Joseph C. Eldrige [sic],[55] who had been appointed an ensign, and was very desirous of being attached to my company.

I knew that a regimental appointment would, in a measure, separate us on duty; but it would place him in a more prominent position and, perhaps, promote his interests. Eldrige was appointed. This occurred on Tuesday, and the Colonel said that my instructions would be ready on Thursday, when he wished me to proceed to Charlotte, on Lake Champlain,[56] open my orders, and proceed as directed therein. On Thursday I buckled all my sword to advance to my station to begin duty as one of the defenders of my country.

[Page 14] I went up the river accompanied by Lieutenants Vale[57] and Curtis[58] attached to my command, and several other officers going to different stations. We arrived at Troy, and took the stage to go north; but at Skenesborough[59] we were obliged to exchange into open wagons on account of the bad state of roads.

We went on at about the rate of one mile per hour until we arrived at Charlotte on the east bank of Lake Champlain. I now opened my orders, and found my headquarters to be at Willsborough,[60] five miles from the west bank of the Lake, which was still frozen over; but the ice was not strong enough to cross on. We stayed with Judge McNeal[61] for two days. The Judge was an "old fashioned" looking old gentleman, and rather amusing. He said that though he had taken no part in the Revolution, some called him Toryish (as he expressed it); but they did not trouble him much. He said that he was not very "cute" at the law, but they made him a judge for want of a better person. On the second night that we were with him, the old judge knocked at my door, and said that I must get up and go down to the Lake to see the ice break up. I did as he desired, hearing a noise like that of hundreds of mills grinding. When we reached the Lake shore we found all the ice in motion; it was still stuck, but jambing and breaking to pieces. The noise continued all night; but in the morning, there was no ice to be seen. Though the wind was blowing down heavily from the mountains on the other side of the Lake, we had the ferrymen prepare their boat to try to pull us across. I was induced to take the helm, and by dint of luffing in the squalls, managed, at length, to get over safely. There was a Clergyman in the boat who had traveled with us from New York. He was much alarmed in crossing the Lake, and in the squalls he clung to my arms — which interfered with my steering. I several times requested him to desist, but he had no command over his muscles, and could not help it. Coming up from New York, he had entered into all our amusements; but he often called us "wicked fellows." On this occasion, I told him that he would be in bad company in case we we should capsize while he was clinging to such a "wicked fellow." After crossing over we had much difficulty in finding horses. At length a fine looking young man said to me, "Sir, I am making a journey, but you may take my horse to Willsborough if you send him back to me immediately I will wait here until the horse is returned." I thanked him, and asked,

[Page 15] "Isn't your name McNeal?" He said "It is." "I thought so," said I, "for I have thus far found the most obliging people in the neighborhood bearing that name." Leaving the other gentlemen to follow, I rode to Willsborough, a little village situated on a small river. It contained a post office, a mill, a forge, a distillery, a tavern, and ten or twelve small houses. I took quarters for myself and officers at "Jones' Hotel"[62] which appeared quite comfortable compared with other things I saw. Vale and Curtis joined me in the course of a day, and it was soon known in the neighborhood that our party has arrived, and many people came to call upon us. Recruits came in even during the first week. I enlisted eight men, a drummer and a fifer. Other officers soon reported for duty, and I established thirteen recruiting stations in my district under Captains and Lieutenants who did not know much about their duty, having just entered the service.

One of the men introduced himself as Captain Trull,[63] and said that he commanded a company. He was a little lame, and I told him that I thought he would find it difficult to make long marches. He replied, "I will show you how I can march," and, holding his cane before his right eye, he stepped off, "dot and carry one," in good style, as he supposed. I, of course, approved. He finally, after two days, went off to his station in good spirits leaving his liquor bill unpaid. It was brought to me, and I found that he had consumed no less than thirty gin slings. I shall have occasion to speak of Captain Trull hereafter; also, of Captain Blachley, from Long Island.[64] The latter made a requisition on me for uniform and side-arms. I told him that the arms and ammunition had not, as yet, arrived; and there was no provision for clothing and arming officers, that they had to supply themselves. He said that it was very hard, and that he would he obliged to go home and sell his little farm to raise the necessary funds.

He said that it was better to do this than to resign, and requested fifteen days' leave, which I gave him. At the expiration of that time he returned armed and equipped; and, being assigned to a station, and receiving his instructions and supplies, he proceeded to duty.

On my arrival at Plattsburgh,[65] I was ordered to take charge of the arsenal and all the batteaux built on the lake. About forty were delivered to me and secured in the small harbor at the mouth of the Saranac River.

[Page 16] Lieutenant Curtis was there, under my orders, recruiting. I found plattsburgh a pleasant station compared with Willsborough. The citizens were much pleased at my coming to take command, and complained that Curtis had annoyed them by marching his recruits about the streets with music at unseasonable hours of the night. This was immediately discontinued on my arrival. I received clothing and supplies from Colonel Clark at Burlington.[66] I had both infantry and artillery recruits, and arms and ammunition for each; tents were erected for the infantry; I took barracks for the artillery, and began building both. Curtis had enlisted a man who was an excellent drummer, but a very hard drinker, and I found it difficult to keep him sober even at drills and parades. I tried many ways, but to no effect. One day directed my first sergeant, George Helmbold,[67] to have one hundred stones as large as a fist collected and placed three feet apart, and a basket three feet from the first stone. I then ordered him to select a man for me to pick them up one by one and place them in the basket, all within forty-five minutes. He considered a moment then (as he had been directed) recommended Jordan, the drummer.[68] It was after morning drill; I ordered Jordan to be called and asked him if he could do this. He said, "Yes Sir; in half the time." He began, and got on well for a time, but not quick enough to do it in the specified time. I said, "If you are behindhand, it must be done again." He now kicked off his shoes. When he finally accomplished his task he threw himself on the ground quite exhausted. I then asked him if he knew why I had ordered it done. He said "no" and I told him that it was for drinking too much. He said that it would cure him, for he had rather be shot than do it again. He had walked nearly six miles, and had stopped to pick up and deposit the stones in the basket two hundred times. It did cure him; at least, he became quite a sober man; and afterwards found this to be an excellent mode of punishment. I remained at this station from June to September,[69] and sent three hundred and thirty well drilled men to Greenbush.[70] Many singular incidents occurred in recruiting. I used to have the men practice firing at the top rail of a fence. On one occasion I noticed that one of them went through all the motions of loading, etc., but that he did not fire. I called him to the front and ordered him to fire off his piece. On **[Page 17]** doing so, he was wheeled about, his shoulder badly bruised, and his musket flew out of his hands. He was never afraid to fire one charge after having fired three all at the same time. The defence of the port and public property being left to me, I organized guards, patrols, counter-signs, etc. Colonel Thorn soon arrived with a regiment of militia.[71]

I furnished the Militia Commandant with the parole and counter-sign, and all went well until the arrival of a large detachment of militia, of which Gen. Mooers took command,[72] but continued to quarter at his house on Cumberland Head, three miles distant.[73] I was, one night, called by the officer of the guard, and, on going to the guard house, found it full of prisoners among whom were Gen. Mooers' aids de camp. I released them and passed them beyond my sentinels. On the next day I was served with an order from the General to strike my tents and remove my men half a mile off, and to desist from giving out parole and counter-sign. I addressed the General in writing and told him that I was placed there on special duty, being ordered by the War Department to take charge of the public buildings, etc., that I should obey his orders under protest, and I requested him to appoint some officer to take charge of the public property and to attend to my other duties, recruiting, etc., as I declined to do further duty except under special order from the Department.

I, however, removed my men to the place assigned. But, in the afternoon, the General called upon me, and wished to come to an understanding. I told him that I presumed that he acted under the article which says, "if the officers of the regular army and the militia meet on the march or in quarters, or on duty, — the one holding the highest commission shall command the whole." He said, "Yes; that he did."

I told him that this rule did not apply in our case, as I was on special duty, and that he was ordered to defend the frontier; that he had nothing to do with my command, and that I had no right to abandon my duty or post. I told him that I deemed a guard etc., necessary to secure the public property, and that he had no guard established. If he would remove his head-quarters to the town and establish a well regulated guard, it would answer the purpose; or if he would authorize an officer in the village to receive the parole and counter-sign, I would furnish them every day. The General acknowledged that he was wrong, and wished me to resume my old station, and proceed with my duties as before. I **[Page 18]** told him that I could not do it informally, as I had obeyed his order, and answered it in writing, and both order and answer were recorded in my orderly book, and it required a written revocation of the order to be likewise recorded. This he sent me and thus the matter ended.[74]

In a few days Adjutant-General Bain, of the British army, arrived with a flag of truce.[75] He wished to see General Dearborn[76] who, he supposed, was at Burlington.[77] He took lodgings at the hotel at which I lodged, and the next morning the militia drummers beat the roll under our windows. Bain saw many of the men, and observed to me that we had some fine troops. I asked him if he had seen my men. He said, that he had seen them at reveille. I told him that they were militia, and that I should drill a small squad of my men at ten o'clock, when he might see them. He did so, and was much pleased. General Mooers informed me that General Bain wished to go to Burlington, and, if General Dearborn was not there to proceed to Albany. I gave it as my opinion that General Bain had more than one object in view — one, to see General Dearborn, and the other to take a small tour through our country to see what were our preparations for war. I suggested that he should not be allowed to go about the country alone, but that an intelligent officer should be sent with him; and that he should not be allowed to stroll about, but keep on the direct stage route. He approved of my views, and sent Major Rathbone,[78] of his command. On the 1st of October the recruiting districts were broken up and the army assembled at greenbush. I was ordered to repair to Greenbush to report to the Adjutant-General with all the men who were with me, who had been enlisted by Infantry officers. A few days before, I had sent off a detachment, and I had but two men remaining who were intended for the infantry. But, with these, I proceeded to Albany.

On the next morning I reported. I had expected to get a company out of the large number of men whom I had enlisted, but I found that they were all scattered through the army as non-commissioned officers at the request of the different Colonels, and I was a Captain without a company. A day or two later, I dined with General Dearborn. He regretted that there was no company of my own drilling for me, but said that Colonel Schuyler would do the best he could for me. When I saw the Colonel, he said that I had best recruit a company for myself, selecting my part of the [Page 19] State as I pleased. I declined on the ground that the army was about to march to the frontier, and that I did not wish to be recruiting while my regiment was in the field. I therefore, offered my resignation, but it was declined, and I finally agreed to pass on, and station officers through the western district to recruit and forward the men to the regiment with which I was to proceed. I had become acquainted with Lieut. Col. Winfield Scott at Albany.[79] He admired the appearance of my two men. He marched a battalion of artillery to the frontier, and we often encamped together during our long march. He often, there-after attended my drills and complimented me highly as a drill officer We arrived at the Flint Hill Encampment[80] together, and joined General Smith's command[81] on the same day. On my arrival at Flint Hill Encampment, three miles east of Buffalo,[82] I was informed that two British brigs,[83] loaded with stores had come down Lake Erie, and were at anchor under the guns of Fort Erie, and that Lieutenant Elliott of the Navy had determined to board them and cut them out.[84] As I had no company, as yet, I rode down to Buffalo, and volunteered to join the expedition. Mr. Elliott regretted that the command of officers was made up. I passed the evening with the party, and accompanied them to the shore, where they embarked in three boats. They rowed up the Lake for some distance, then, dropping down with the current, they boarded the brigs. One was lightly armed and made some resistance, but was soon overpowered, the cables cut, and got under weigh for the American shore. One was brought in and anchored under a battery manned by some three hundred militia under command of Major Miller,[85] as I afterward learned. The other brig drifted down and grounded on the west side of Squaw Island,[86] within six hundred yards of the British shore. As soon as the American brigs, Adams and Caledonia, got under weigh, my servant, William Williams,[87] and I rode down the beach to see them beat in. The road turned up the hill to the battery and I discovered that the battery on the opposite side had opened its fire upon the Adams. I inquired in vain for the commanding officer of the militia battery. I asked the men for ammunition and found all excepting port-fires. The men were stooping under the parapet at every flash, and none would assist me to return the fire. I then sent Williams out on the road to invite in any stray soldiers or sailors whom he could meet; he soon returned with three or four. We made a fire of rails on the parade, the men loaded and **[Page 20]** I sighted the pieces, and a man with a burning rail touched them off, as I directed. We were soon out of ammunition; there was firing below us on the bank, and I sent my man Williams to endeavor or to get some nine pounder ammunition. He found Colonel Scott in command, who, knowing my man, sent me a good supply. So we kept up the fire until the prizes were secured and the cargo landed, when the British batteries ceased firing. When the prisoners were landed, I took charge of them, procuring a few of the militia to act as guard, and proceeded with them to Buffalo. I was much fatigued, and several times during the march I found myself sleeping on my horse. I delivered my prisoners to General Hall,[88] and proceeded to my regiment which was then on the march from camp to go to the protection of the brigs, particularly the Caledonia which was aground on the island. A strong detachment was placed on Squaw Island, which repelled and severely cut up two or three parties that attempted to board the brig. Before evening, the 56th and 13th Regiments of Infantry,[89] Scott's battalion of artillery and Colonel Stanton's[90] and Colonel Mead's[91] regiments of militia were assembled at the mouth of the Conjacaty Creek expecting a visit from the: British. Colonel Schuyler was in command. In the evening, Scott's servants having foraged and found some beef steaks, he invited Col. Schuyler and several other officers, among them, myself, to partake and, while eating, the militia guards reported that the river was full of boats crossing above and below.

The Colonel ordered me to take a few men and get a good position above, and report frequently what I saw. I sent frequent messages that nothing was stirring in that quarter, but at about two o'clock, I heard troops moving on the road. I challenged them and was answered by one of our officers "We are on the retreat, get your horse and join us." I had left my horse at General Porter's[92] and, proceeding there, I found all quiet. Elliott[93] was also there. We stood on the piazza and saw the troops passing. I challenged, and was answered by Captain Sprowell, of the 13th, 2nd Company;[94] he and Martin joined us.[95] None of us believed the enemy to be near, and if it was, it was our duty to fight and not retreat. Colonel Schuyler was very censurable. Our two companies halted at the battery at which I had been in the morning, and determined to stand and meet any enemy that might advance; but my command was soon put to an end by the arrival of Major **[Page 21]** Huyck of our regiment. He requested me to mount his horse (mine having strayed) and ride to Buffalo to alarm General Hull's command. When I arrived there all was quiet. I aroused General Hall, who was much bewildered, and did not know what to do. At my suggestion he called a trumpeter who, after blowing for some time, got together some thirty horsemen without arms, having lent theirs to the boarding party. Thus ended this disgraceful affair, as a beginning to the war. To the honor of Lieut. Col. Scott, I may say that he did not obey the order to retreat, but remained on the ground all night.

On the next day we all assembled again at Flint Hill Encampment. Colonel Schuyler sent for me to advise him as to his best course, whether he should arrest or challenge an officer who had said that he acted in a cowardly manner by ordering the retreat. I advised him to do neither, but to let it pass. He remained but a short time with the regiment: he was an excellent disciplinarian and drill officer. He was soon Adjutant-General of the command of his brother-in-law, Colonel Cushing.[96] We passed the winter of 1812-1813 camping at Flint Hills, and marching to Buffalo and Black Rock and back.[97] The troops suffered very much for want of provisions and clothing suitable to the climate. We built comfortable log barracks, but we lost by death through the winter some three hundred men. My company, and that of Captain Morgan[98] marched to and from Buffalo twenty times during the winter. Old Colonel Porter, known as "Blow hard,"[99] commanded. He frequently dreamed that there was an attack on Buffalo, and we were ordered to march from comfortable cantonments at Williamsville, eleven miles from Buffalo.[100] And when there after a march, the tracks of which could be traced by the blood from the feet of the men (for they were nearly all bare footed) we were compelled to encamp in the streets being unable to procure quarters.

Early in April, the whole army was marched to the river to perform drill in order, first, to keep the militia from rebellion, and second, to storm the batteries on the British side of Niagara.[101] Both of which were effected. A brigade of Pennsylvania Militia was mutinous, and even threatened to take General Smith, our Commander, out of camp.[102] We formed line of battle, on one occasion, expecting an attack from them. We soon left Camp Rock, and returned to the cantonment at Williamsville. We remained there [Page 22] for a few days, then the whole force marched to Black Rock across the Niagara River and stormed the batteries on the British side. I was ordered to general Smith's quarters where I found a number of officers assembled.

The General stated that we were to cross in three divisions; at twelve o'clock, and storm the batteries on the margin of the river. He asked me now many men I reported for duty. I answered, eighty.

He said he would give me seventy more and and additional subaltern, and I was to have three boats. The General remarked, "Virginia will not complain tonight." Most of the officers present were Virginians, and he thought me to be one also by my name. We were dismissed and we found our respective commands; but the boats were full of ice and water, and deficient in oars. It was near daylight before we-moved. We stormed the batteries with a small loss, and took a number of prisoners; but the enemy having received a large reenforcement from Fort Erie, we were compelled to recross the river leaving a small detachment behind, which was captured.

Among those taken was my first sergeant, George Helmbold, who had separated from my commend in the bustle of retreat. He had a presentiment that he should he killed in the first engagement, and he told me afterwards, that finding, after a few shots had been exchanged, that he still lived, he gave up his superstition, and was ever after a brave man. After our return, General Winder was ordered to cross with his brigade,[103] but the fire was so brisk that he could not effect a landing. The prisoners taken from us were marched to Fort George, and examined by Colonel Myers of the British Army.[104] Helmbold had held prisoner, during the night, a British sergeant, and imprudently, as well as unmilitarily, had taken his sword and belt. The sergeant was missing, and it was reported that Helmbold had killed him, while a prisoner, and taken his sword and belt. The Colonel asked him his regiment, and the name of his Captain. Helmbold said that he belonged to the 13th Regiment, and Captain Myers' Company. The Colonel enquired particularly about the size, complexion, and appearance of his commander, and then said, "I know your captain; he is a British deserter, and I should like to see him." The sergeant replied, "Perhaps, sir, you may meet him by and by;" and so he did, for at the taking of Newark and Fort George, Colonel Myers [Page 23] was wounded and taken prisoner. A few days after the battle, in company with some of the officers, I visited the British prisoners. The moment my name was mentioned, Colonel Myers related the above, and said, "I expressed a wish to see you, but not under the present circumstances; I am mistaken, you are not the deserter." Then, he said, "Your sergeant is a brave fellow; he would have been condemned to be shot, had it not been for the wife of the missing British sergeant who, hearing the charges against Helmbold, came down to Fort George to say that her husband was wounded and taken to a hut, where he then was, but that he had not been ill used by Sergeant Helmbold further than having been deprived of his side-arms and belt. This saved Helmbold's life.

The expedition thus ended, we marched to our cantonments at Williamsville.[(105)]

Our next attempt was to cross and take Fort Erie. We started one morning just after daylight and rode up the river from Conjacaty Creek to Black Rock.

An officer was sent over with a flag to demand the surrender of the fort. If it was refused, he was to make a signal to that effect on leaving the shore. This, he omitted to do, and, much to the dissatisfaction of all, we were ordered to return and secure the boats. The plan of the first Campaign of 1812 was as follows: — The army was formed with the right on Lake Champlain, under General Pike;[(106)] the center, on the Niagara, under General Smith; the left at Detroit under General Hull,[(107)] and the whole under General Dearborn. The left wing, under Hull, was defeated by the command of Gen. Proctor.[(108)] This occurred through cowardice or treason; from the mode of attacking the British fort, the latter was strongly suspected. No general expecting resistance would attack a fort in column, when he could do so in line; for a ball striking the head of a column would completely rake it, which would not be the case in line.

I conversed with Col. Miller,[(109)] Col. Snelling,[(110)] and Col. Watson;[(111)] they were all satisfied that the affair was arranged between Hull and proctor. From the loss of the left wing, the center became the most important division. This was to push forward and drive down the British forces from the protection of Kingston[(112)] and Montreal;[(113)] and the right wing was to join them at St. Regis on the St. Lawrence;[(114)] and the whole body was then to move on Montreal. At the same time three squadrons of armed vessels were organized [Page 24] and equipped; — one on Lake Champlain under command of Commodore McDonough;[(115)] one on Lake Ontario under Commodore Jones,[(116)] and one on Lake Erie under the command of Commodore Perry,[(117)] and all under the command of Commodore Chauncey. They were to co-operate with the armies. The left wing having been disposed of by General Proctor, he remained unopposed until forces took the field under Gen. Shelby[118] and Gen. Harrison. They were successful against Proctor; Tecumseh, the Indian Chief, was killed by Col. R. M. Johnson, and the whole British and Indian forces were beaten.[119] Commodore Perry was also successful on Lake Erie, taking the whole British Squadron; McDonough was, in like manner, successful on Lake Champlain; while Chauncey[120] was effectually co-operating with the army of the center, on Lake Ontario, and holding in check the British Squadron. Early in April Gen. Dearborn and Commodore Chauncey arranged for the invasion of Canada. The whole land force, excepting a detachment then at Sacketts Harbor,[121] assembled at and near Fort Niagara, embarked in Chauncey's Squadron and made a successful attack on little York.[122] We there lost our brave General Pike. I must now digress a little to give an account of the battle of Queenstown,[123] which I omitted to bring in its proper place. The second battalion, as I have before stated, had taken some boats up Lake Ontario to Fort Niagara; while the first battalion marched on and encamped at Flint Hills, three miles from Buffalo.

Our troops at Fort Niagara, and small detachments from regiments, with Col. Mead[124] and Col. Stanahan,[125] under Gen. Wordsworth[126] prepared under Col. Philip Van Rensselaer[127] to attack Queenstown. Gen Stephen Van Rensselaer, one of the best men of the age, though not the greatest general, was ordered to the Niagara frontier in command of the militia. His nephew, Col. Philip Van Rensselaer,[128] a brave officer of some experience accompanied the General, who now determined to collect his forces and select a night in which to make a descent. General Smith's brigade, then at Flint Hills, and the troops at Fort Niagara were ordered to assemble at Lewiston,[129] in accordance with orders from general Van Rensselaer, who ranked all the other officers on the frontier at that time. We took the line of march and proceeded toward the rendezvous; but when we were within seven miles of our destination the order was countermanded and we returned to camp that night after a march of twenty-eight miles over the worst possible roads.

[Page 25] That day I lost, through sheer fatigue, a fine, active young man; he was a sergeant named Comstock.[130] We left him at a house on the road; but coming into camp, as we-had promised, he was sent for, but found — dead. Two days after this we received another order to proceed to Lewiston, and on the way an express met us with orders for us to hurry on as the troops taken from Fort Niagara, including the second battalion of the 13th Infantry under Lieut.-Col. Christie,[131] and a body of militia under Col Mead and Col. Stanahan, commanded by Gen. Wordsworth, had crossed over to Queenstown in the night, and, after a sharp struggle, had taken possession of the town and heights. We moved on with great rapidity, but before we reached Queenstown our troops were overpowered, the British having received reinforcements from Fort Erie and Fort George, while many of our militia at Lewiston refused to go over.

The fight was an irregular one; most of our officers and many of our men were killed or wounded in crossing the river or soon after landing. Among the wounded were Col. Solomon Van Rensselaer, Lieut.-Col. W. Scott, Lieut.-Col. Christie, Captains Lawrence,[132] Malcom,[133] Wool,[134] Armstrong,[135] of the 13th Infantry, and several others of the 13th were killed. This severe loss among the officers arose from the fact that all those who could raise a few men took them over in separate commands. Captain Wool and Captain Ogilvie[136] of the 13th Infantry both claimed great honor for having ascended the heights and taken an eighteen pounder that was mounted on a pivot. The piece was taken, but it has always been uncertain which of the officers took it. We arrived at Lewiston a few hours after the surrender; many of our wounded men had got to the boats and returned. The prisoners were marched to Fort George and thence sent to Montreal. The whole affair ended disastrously from mismanagement; for had Smith's brigade of about fifteen hundred men been ordered up in time to join the expedition not only Queenstown, but Fort Erie and Fort George would have been taken, and we should, undoubtedly. have been in possession of the British shore from Lake Erie to Lake Ontario; the British could not have withstood our united forces. Immediately on our arrival at Lewiston,

General Van Rensselaer gave over the command to General Smith. I was ordered to enroll the remainder of the 2nd battalion of the 13th Infantry, to collect them, provide for their wants, and command them as a detachment. I found them [**Page 26**] in a most wretched condition, in barns, under sheds and hay-stacks. I drew tents, and collected the men, provided medical assistance for those who were severely wounded, and from the remains of those four companies, I formed my grenadier company of the 13th Regiment, and for appearance, bravery, and knowledge of their duty they were not excelled in our army or in that of the enemy. We remained at Lewiston for ten or twelve days, and then marched hack to our old quarters at Flint Hills where the wounded could he better cared for. We remained here until ordered to build barracks at Williamsville, on the "Eleven Mile" Creek. The cantonments were built by the troops, who marched and encamped there. The various parts of the work were systematically given to different parties, each under command of a captain. The duty of myself and party was to procure and haul sawed lumber.

The barracks were, at length, completed, and were very comfortable. During the winter (1812-13), we were-on the defensive, and, with the exception of many alarms, and the frequent marching of my company (1st or Grenadier of the 13th) and that of Captain Morgan of the 12th to Buffalo and back, little or nothing worthy of note occurred. Marching to Buffalo so frequently was very severe duty. I occupied my quarters only one night after their completion; the remainder of the time I quartered in a tent either at Buffalo, or at the Cantonment. I often encamped in the street in Buffalo, there being no quarters to be had. Finally I secured the Ball-room[137] for my men and a shoemaker's shop[138] by way of quarters for myself. Here, during the winter, many occurrences took place which were very amusing at the time, but they would not be interesting to read at this late period. As I before stated, the campaign of 1813 began with the affair of Little York.

The troops remained there only two days. They took one or two schooners with stores, etc., burned a vessel on the stocks, and re-embarked and stood over for Fort Niagara.

Here the dead were buried and the wounded taken care of as well as circumstances would admit. Our squadron left Sackett's Harbor. General Smith's brigade, then at Williamsville encampment, was ordered to march to Fort Niagara. We descended the river in bateaux as far as the old French fort, one mile from Niagara Falls and just opposite to Chippeway.[139] I was detached with two hundred men to take the boats back about two miles. We [Page 27] accomplished this with great difficulty, the current being so strong that we could barely stem it. After effecting our object, we set out on our return to Chippeway. The men shouldered their oars for convenience in carrying them. When we approached the garrison, the troops, seeing a body of men carrying arms fourteen feet long, feared an attack and beat to arms; but our long weapons were harmless. At reveille we proceeded on our march to Lewiston where we halted. There an order reached us directing General Smith to re-enforce Fort Niagara with our regiment of infantry; for an attack was expected from the enemy's troops at Fort George by way of retaliation for the affair of Little Rock. We were just pitching our tents to shelter our men from a snow storm, when the order came. Colonel Christie, who was in command of our regiment, asked me if the regiment could move on immediately. Some of the other captains were consulted, and we finally started on the march without baggage.

The night was very dark, cold and stormy, and the roads were as bad as can be imagined. We commenced the march, left in front; so my company brought up the rear, a very unpleasant post, as those in the rear cannot go on until all stragglers have passed, when it is sometimes very difficult to catch up with the front. On arriving at the salt battery, Lieut. Col. Schuyler informed me that the field officers would go immediately to the fort, that I must halt and get shelter for the men, and, if it was necessary for us to come on, a rocket would be fired, in which case we must move down as quickly as possible. He gave me the parole and countersign. Our men took shelter such as they could find.

Captain Archer,[140] of the light artillery, who was stationed in the battery, invited the officers to his quarters and ordered refreshments; but in less than half an hour the signal rocket was seen at the fort. I immediately formed as many men as I could collect, and left Captain Martin[141] in command to follow as soon as possible, and to leave the same order to Sprowl[142] and the other captains. We groped along through Egyptian darkness, the rain fell in torrents, and the mud was ankle deep. When I arrived at the chain of sentinels I was obliged to advance slowly through the mud, slipping down now and then among the stumps and stones.

I answered the challenge; "a detachment of the 13th Infantry;" the answer was the discharge of a musket, and the-fire was repeated all along the line of sentinels.

[Page 28] After some difficulty exchanging countersign, etc., we gained admittance to the fort and found the garrison under arms; as they had thought, at first, that we were the enemy. Everything went on quietly for the remainder of the night, and at daylight the rear of our regiment arrived. We had no quarters, but tents were furnished and we encamped on the parade. The next night the wounded from Little York were landed, and the fort was then like a hospital. The mess house and the stairs leading to the mess room were all filled. We remained on duty in the fort for several days, and then the whole force, excepting the regular garrison, encamped on Snake Island.[143] It was literally Snake Island, for it was full of snakes, The men frequently shook rattle snakes from their blankets in the morning, and at every drill or parade snakes were killed on the ground.

The inhabitants of the island said that the snakes came down from the mountains in the spring to get water at the lakes; but that before they reached the water they were harmless. None of the men were bitten by them, at all events. While at this encampment I was engaged as second in an affair of honor between Dr. James Bronaugh[144] and Major Stonard.[145] Major Z. Taylor (late President of the U. S.)[146] was associated with me as friends to Major Stonard. Captain Randolph[147] and another gentleman were friends to the Doctor. The Major was shot in the thigh. He died of the wound in May, 1813. (See Appendix.)

While at Snake Hill [ed. note: Myers meant Snake Island][148] every preparation was made for taking Fort George. It was arranged that on the night before the landing, Fort Niagara, the salt battery, and a battery which I built one night assisted by Lieutenant Totten as engineer (now General Totten, U. S. A.)[149] should open fire of hot shot on Fort George and the batteries on that side, and at four o'clock, A. M., all the troops were to embark and wheel out by regiments. Lieut.-Col. Winfield Scott commanded the left with his light artillery supported by infantry.

Next, on the right, was the 13th Regiment, commanded by Lieut.-Col. Christie; and the other regiments according to their rank. Each regiment formed a line-of boats and moved in silence at wheeling distance.

The companies were so arranged in each boat as to form line immediately on landing. It was a misty morning, and our fleet had taken up its position under cover of the fog to effect our landing. The cannonading on the right was very brisk and effectual; all the **[Page 29]** wooden buildings in Fort George were soon in flames, and the garrison marched out and formed on the plain ready to move in any direction. We had about five miles to row to the place of landing; as we neared the fleet the fog lifted, and a view was presented of our fleet at its moorings, and our full force approaching with flags displayed and our bands playing "Yankee Doodle." The garrisons of Forts George, Erie, and Chippeway, which had joined in the night, were in line near the bank. The firing now opened from the shore and shipping.

Lieutenant Trent,[150] in the schooner "Julia," with one twenty-four pounder on a pivot, took a station opposite to a battery, furled his sails and moored his schooner for a "regular set to." Scott effected a landing, but did not succeed in advancing until re-enforced. I commanded the in-shore boats of our regiment, and was rowing for my position when General Boyd, who was on shore, called out, "Wheel in my jolly snorters," (the sobriquet by which our regiment was known). I immediately wheeled my boat when an order was passed from boat to boat "to keep on." I answered in the same way "Superior orders from shore." I landed and the remainder of the regiment followed, and Scott with all who were with him mounted the bank together. The contest was sharp, and soon ended. The British retreated, and were soon on a rapid march for "Twelve Mile" Creek.[151] We advanced to Newark, but were annoyed by a fire of hot shot from a thirty-two pounder on a pivot in barbette in a block house. Colonel Christie was ordered to detach a force from his right and take it. I was ordered to take two companies from the right, and Lieutenant McDonough[152] was to accompany me. Lieutenant Paige[153] commanded my second platoon or company.

When I approached the block house, I detached him to the rear to take prisoners if they attempted to jump over the parapet; for it was a battery on the water side, and a block house on the land. I entered through a sally-port five feet high and three feet wide. The garrison jumped over the parapet and ran. Paige had not gained his position, and I never saw Mr. McDonough or his twelve pounder. After taking this position, the army found nothing to obstruct its progress. There were a few reports to the effect that large quantities of ammunition had been placed in the public stores with trains of powder leading to them, but examinations proved them to be untrue. The whole body advanced as far as the public **[Page 30]** stores near the house of Mr. Black.[154] Here the main body halted, and Lieut.-Col. Scott proceeded with his detachment to the ruins of Fort George. He found no troops to oppose him; all were on the march to the "Twelve Mile" Creek. The fired buildings were still burning. I was detached for main guard. The park of artillery was in the center, and the other troops were at the right and left facing south — the right near the road leading to the batteries. I was posted as main guard at about five hundred yards south of our line, and exactly in front of the park of artillery. I had occasion to go to Mr. Blacks to instruct the family how far they could move without interruption. I took a glass of porter and some crackers and cheese with Mr. Black.

At about eight o'clock, Col. H. V. Milton,[155] as officer of the day, visited me and requested me to have fires made up. I complied, though to have fires on out post in the enemy's country was contrary to what I had been taught. He said that in case of an alarm I might withdraw my guards and place them between the fires and the line. At about eleven o'clock the sentinel on night guard fired. I sent a non-commissioned officer and a file of men to enquire the cause. It was reported to be a false alarm. I had retired as directed. The officer of the day then rode up and said that I had been in great danger. The artillery, supposing my guard to be the enemy had aprons off guns, the guns primed, and port-fire ready; and had he been five seconds later my guard would have been cut down by grape shot from the whole park. Thus, by the providence of God we were saved. There was no alarm during the remainder of the night.

There was no necessity for stopping at Newark except the Generals having been hungry, and Mrs. Black having had a good dinner. I suppose they thought they had glory enough for one day without following up the enemy. The same night a division under Colonel Burns went down, crossed at "Five Mile" Meadows and took Fort Erie without resistance, and the next day marched down, took Chippeway, and joined us.[156] The center was now concentrated at Fort George, having possession of everything from Fort Erie, on Lake Erie, to Fort George, on Lake Ontario. Our camp was for many days full of Canadians, who came in for protection. After giving the enemy time to reach Burlington Heights, at the head of the lake, a strong detachment of our army under General Lewis marched in pursuit.[157] They halted on the first night at "Twelve Mile" Creek.

[Page 31] Our detachment under General Lewis marched from Fort George without baggage on the first day of June, 1813. It consisted of the 12th, 13th, and 14th Regiments of Infantry, Colonel Burns' regiment of Dragoons, and a strong detachment of Artillery, numbering in all, rank and file, about three-thousand men. Our object was to attack the British in their entrenchments at Burlington Heights.

On the 2nd we halted to give the men a little rest at Crooks,[158] twenty-two miles from the British lines. We had had a hard march of two days over bad roads obstructed by broken bridges and fallen trees placed in our way by the retreating enemy. Many of the officers took breakfast with Mr. Cook,[159] wealthy gentleman; and while I was enjoying the luxury of putting on a clean shirt, kindly lent me by Dr. J. Bronaugh with the injunction not to tell where I procured it, a party of Indians took possession of a high and almost perpendicular ridge of rocks in front of the house and commenced a heavy fire on the dragoon horses that were picketed in the door yard. Lieut. Joseph C. Eldridge, of my company, and adjutant of the regiment with the camp guard accomplished immediately; but with great difficulty, what seemed to be impracticable — he scaled the face of the rocks. At the same time, I was ordered to march to the rear of their retreat; but before I reached the top of the ridge they were all dispersed by Eldridge and had escaped.

We continued our march through mud and mire, it sometimes required six teams to drag our baggage wagons and artillery; the day was extremely warm, and we were obliged to repair several bridges before we could cross the streams. We halted in the evening at "Forty Mile" Creek.[160] A detachment of boats with provisions and ammunition had followed our march. They were discovered by Commodore _____ and we dispatched the schooners to destroy them.[161]

As soon as we halted, Captain Archie of the artillery[162] was ordered to the lake shore, supported by the 13th. Furnaces were erected and hot shot was soon prepared and used to such effect as to drive off the enemy. The remainder of our forces were bivouacked [sic] on either side of the road leading to the position of the enemy. Lieut. Van V_____[163] commanded an advanced picket guard half a mile in advance on the road. This was the evening of the 3rd of June, and the 4th being the birthday of King George III, the enemy felt in good spirits, no doubt, .and determined to precipitate our attack.

[Page 32] They moved their full force towards us in silence, bayoneted the advance sentinel, found the officer of the guard asleep, took the guard and marched straight into camp before they were discovered. Great confusion, of course, ensued. General Winder and General Chandler were taken prisoners[164] while giving orders to the enemy, believing them to be their own troops, in consequence of one of our regiments having changed its position in the night to get better encamping ground. As soon as the firing was heard, Col. Christie moved our regiment along the beach with the intention to get in the rear of the enemy, take possession of the bridge over "Forty Mile" Creek, and cut off the retreat. We had nearly reached our point when an express came with orders to halt, and for the field officers to attend a council of war. General Lewis had been called back to Fort George, Gen. Winder and Gen. Chandler were prisoners and Christie thought that he ranked all the other officers. But, on going to the council, he found, much to his chagrin and mortification, that Col. Burns ranked by date of commission.

The question now was, shall we pursue or fall back? Our regiment was counter-marched over the field of battle after having taken twenty or thirty prisoners. When we arrived at the ground we found it strewed with dead and wounded of both parties to the number of four or five hundred. The troops on both sides were scattered. We buried the dead and stacked and burned the arms and baggage for want of transportation. We brought off the wounded and prepared to follow our retiring forces. I was informed that the British General S____ [165] was lying among the dead. I sent an officer in search of him; he found a blanket and one of the General's pistols, but, like Lord G____ in the play, the General had carried off his own corpse. [166] At noon we started after our retiring troops, and joined them at the "Forty Mile" Creek. Col. Schuyler asked permission to follow the retreating enemy with our regiment, but the Council of War would not consent.

If it had done so we might have retaken our two generals, and, perhaps, many prisoners, for General Winder told me, when he returned on parole, that at twelve o'clock the enemy had not collected five hundred men; they were not three miles from us, and they expected an attack every moment. [167] On the 5th we continued our march and on the 6th we arrived at Fort George, hav-[Page 33]ing gained no laurels and having lost nearly three hundred. We were now on the defensive, and General de R____ soon after arrived and took command. [168] We now fortified our camp with a ditch in front, Fort George on the right, a strong battery on the left, and the river in the rear. About one-third of the men were detailed for fatigue duty to diminish the size of Fort George, which was so large as to require a garrison of seven hundred men to defend it properly. The men for duty were drilled eight hours per day. Attacks on our picket guards were frequently made by parties of British and Indians. It occurred to me that a six pounder would be of great benefit to each of our picket guards. I tried its effect for the first time on picket number two, on the swamp road.

I had it loaded with grape, and at four o'clock in the morning an attack was made. I turned out the guard, masked the piece, and, when all was ready, opened by half wheel by sections and touched off the gun.

The Indians set up a shout like that of a thousand devils, and were off. This plan was generally adopted. Our sentinels were frequently shot on post. One day I placed my whole chain each under cover of a rock, tree, or other object so that they might see without being seen. This, also, was afterwards generally adopted.

The Indians often approached our line of sentinels in the night and waited for the grand rounds. When the sentinels challenged, the Indians, guided by the sound of the voice, fired, and some-times hit a man. One day I drilled my guard to strike once on the cartridge box when they heard an approach instead of challeng-ing, and I instructed the rounds, relief, and grand rounds when they heard a sentinel strike his cartridge box once to answer by striking twice and to advance. This plan was followed, and, no doubt, saved many lives.[169]

We were in the habit of reconnoitreing [sic] with small detach-ments, and we had many little fights with the picket guards. The enemy did the same, and one morning an attack was made on our picket number three near Butler's farm. A detachment of forty from our regiment was ordered out. Adjutant Eldridge was form-ing it and he asked me how many men he should take from my company. I told him as many as he pleased, and he took twenty-nine men. He had promised me not to take such commands as it was not pleasing to the other subalterns to be deprived of them by the adjutant, whose duty it was [Page 34] to act only with the regiment; and thus give others a chance to signalize themselves. I repeated this while they were forming; he said, "Only this once, Captain, and never again; may I take your pistols and belt?" I said "yes," and he marched. After a time, the firing was very heavy. I went to the Adjt.-General for permission to go out with the remainder of my company.

He answered, "An order has been given for your regiment to march." Though we marched rapidly, we were too late to prevent the horrible massacre of our brave young officer and his men.

On his arrival at the picket, he had found the guard engaged with a very superior force. He joined in the action; the enemy fell back, and he pursued into a morass surrounded by thick, bushy woods. There, he was surrounded by a large force of British and Indians and cut all to pieces.

Only three out of the forty taken from our regiment, and but few of the picket guard returned to camp. Our regiment passed the mangled dead in pursuit of the enemy, which had fled with the scalps. The bodies were stripped and horribly mutilated. Lieut. Joseph C. Eldridge, the Adjutant of the 13th Regiment, who was among the killed, was a young man of great promise, and much beloved by his fellow officers. [170]

Being recalled from the pursuit, we had to perform the melancholy duty of burying our mangled officers and men in one common grave.

The first battalion of our regiment was ordered to Queenstown to accomplish two objects. First, to send a captain's command at reveille the next morning to reconnoitre in the neighborhood of Lundy's Lane and Niagara Falls,[171] to see Mrs. Wilson at the Falls,[172] and to obtain from her all the information possible respecting the strength of the enemy at, or near St. David's,[173] the number of Indians employed and acting with the British, and such other information as she might possess. Also, to collect and forward by the river all the public property found at Queenstown. I had been unwell for a few days, and unfit for duty; but Major Huyck,[174] who was in command, was so desirous of having me go with him, that I consented, and waived my privilege of being on the sick list. Sometimes I marched on foot, and at others the Major dismounted and lent me his horse.

We arrived in the night, and bivouacked [sic] in front of the Hamilton House, then an elegant establishment,[175] on the bank of the river, [Page 35] but now almost a ruin. The Major opened his orders and directed me to command the reconnoitreing [sic] detachment. I sent a non-commissioned officer and a file of men to impress a horse, and they had just returned with a fine one, and I was forming my detachment of one hundred and twenty infantry and sixteen riflemen, who accompanied us from the fort, to act as flankers, when Major Chapin arrived with forty mounted Canadian Volunteers.[176]

I knew the Major to be a great humbug, and I told Major Huyck that I must relinquish the command, as I could not march under the orders of Major Chapin of the militia.

Chapin said he did not expect to take command, but would act under my orders. We then moved on about four miles. Then I detached the Major with his volunteers to take a back road and join me in Lundy's Lane, which he did. We proceeded to the Falls.

I had an interview with Mrs. Wilson, an American lady who married an Englishman. Their house was a great rendezvous for the British officers. She reported fourteen hundred regulars at and near St. David's, about seven miles distant, and a body of seven hundred Indians from Lake Superior. She said that Captain McGilvery,[177] with a detachment of Indians and regulars, had been down that morning. He knew of our detachment being out, and would attack us on our return at a pass cut out between two hills, about four miles from Queenstown. She feared that his force and the position he had taken would render my return almost impossible. Having obtained all the information I could gather I marched for Queenstown and — a fight.

I sent twenty men, infantry, as an advance guard, and four of Chapin's Volunteers as a vidette [sic][178] in front of the advanced guard, then my column of one hundred infantry, with eight riflemen as flankers on the right and left, and Major Chapin with thirty-six mounted volunteers to bring up the rear. During the march, Mrs. Wilson's report of the intended attack was confirmed by a young lady, the daughter of a British surgeon,[179] whose house we had surrounded on account of information received that it was the residence of a British officer. He appeared in uniform and delivered his sword, saying that he was my prisoner; but his uniform satisfied me that he was not. I, however, asked him his rank, and he said, "I have the honor to be Hospital Surgeon in His Britannic Majesty's Service.["] I, of course, returned his sword, as he was a [Page 36] non-combatant, and not a prisoner. The daughter was very pretty, and quite communicative.

She wished me to dismount and take refreshments, which I declined; she, however, brought me some strawberries and cream. She said she feared I would be killed or made prisoner; and she hoped, if I were made prisoner, that I would take a parole and return there; I had been polite to her father, and it would give them great pleasure to make me comfortable should I fall into the hands of their people. We proceeded, and just as the vidette had entered the descending grounds, the advance guard following, the enemy opened a heavy fire from both sides of the road. They should have reserved their fire until the main body entered the deep cut, when their advantage would have been very great; but, as yet, the main body was on higher ground than even themselves. I threw the column into the right and left, making a half wheel from the center of the platoons, and gave them a well directed and deadly fire to the right and left. They stood only three or four discharges, when the Indians on the right of the road gave way with a tremendous whoop and yell, and the British, on the left, broke, threw off their accouterments and ran through the woods. We passed the ravine and, procuring spades from a farm house near, we buried three of our men killed. Major Chapin, with his thirty-six Canadian Volunteers, now came up. He said that he had stopped at a house where, he was informed, some British officers and men were concealed. We returned to Queenstown.

Major Huyck met and embraced me in the street. He had heard heavy firing and feared that I had been cut off. He had sent all but eighty of his men to the fort with public stores. He had his party formed, and expected an attack every moment; in fact, when he saw us approach, he believed us to be the enemy, and was prepared to resist. We marched four miles together towards Fort George, and halted for the night. Some deserters came in and said that we had handled them very roughly, and killed many of their men.

The next day, I reported to General Dearborn the result of my reconnoissance [sic]. Upon being asked my opinion, I stated that a proper force to make an attack would be a thousand infantry, well supported by artillery and cavalry. Colonel Butler [sic: Boerstler][180] having heard of my report, volunteered three days after to march his regiment over the same route. His offer was accepted, and he marched [Page 37] with seven hundred infantry and a detachment of artillery. He was attacked, on a woody road, by a considerable force; he sent an express for re-enforcement, and our regiment immediately made a forced march to Queenstown. There, we received notice that Butler [Boerstler] with his whole force were prisoners. Our regiment alone being too weak to follow the enemy, unsupported by artillery or cavalry, returned to the fort, and the boasting Colonel and his command remained prisoners. At this time Commodore Perry made a requisition for a company of infantry to act as marines. When I heard of it, I went immediately to headquarters to offer my services; but I was just a few minutes too late to be accepted, and thus lost the opportunity of being present at Perry's victory on Lake Erie, when he so signally defeated a superior force and gained so much glory. This, with Harrison 's victory,[181] destroyed the British power in the West, and the way was now clear for a descent upon Kingston and Montreal.

It was arranged that a strong militia force should be collected at Fort George to permanently secure the Niagara. The whole regular force in upper Canada was to consolidate at Sackett's Harbor under General Wilkinson.[182] General Hampton's command,[183] then at Chateauguay, "Four Corners," in Clinton County, New York,[184] was to join us with four thousand men and provisions for the whole army at the mouth of Salmon River on the St. Lawrence.

Every arrangement was made by General Wilkinson at Fort George. The army embarked in some six hundred craft of different kind. Each division and regiment displayed a different flag. The commander of the boats had descriptive lists of each flag, so that they might know one another. The movement was made under convoy of the fleet commanded by Commodore Chauncey.

We left General McClear [sic: McClean],[185] of the State Militia, with his brigade. He remained but a short time when he disgraced our cause by burning the village of Newark. He did it under the plea that he could not defend it, and that, if left, it would afford winter quarters for the enemy. This act, so uncalled for, brought on a severe retaliation. Through either the cowardice or treason of Captain Leonard,[186] the British took Fort Niagara without meeting the slightest resistance [*ed. note: this is a significantly inaccurate statement; after the initial surprise of the attack's beginning, there definitley was armed resistance, and there were combat casualties on both sides*]. They then, at different points, crossed the Niagara with strong detachments, and burned Youngstown, Lewiston, the Tuscarora Indian Village, Black Rock, and Buffalo;[187] killing many, and destroying an immense amount of property. We proceeded [**Page 38**] down the lake in continual gales, making ports for shelter as often as possible. Sometimes, we would land on the beach where no inlet was to be seen; a hundred men with spades would soon remove a bar of sand thrown up by the surf, and open a passage to a commodious bay.

We were much scattered during the descent; many of our boats were stranded; mine, with one hundred and thirty men and women, was among the number, but all were saved. Our pilot declared that our boat could not live in such a storm, and that he could make a safe port; but that, if I determined to keep on, the whole responsibility must rest on me.

I, therefore, directed him to bear away for a port. We saw lights on shore, boats having landed there and left their fires burning. When nearly in I told him that I heard the breakers, and asked him if he could haul off; he said it was impossible. We then kept off and ran on; the men were ordered to sling their knapsacks and stand ready to jump out on either side when the boat struck, and endeavor to shove her up. I stood on the bow with the painter , and, as she struck, we all jumped out; the breakers went over us several feet; the third time she struck, she went to pieces and sixty barrels of provisions floated off. We all gained the shore, a narrow sand beach and an abrupt rocky bank at least one hundred feet high. We found that we had landed at Mexico Bay.[188] We collected some wet provisions along shore, and some sails which I ordered to be so disposed as to afford shelter for the men; then I started along the beach to explore.

About two miles away, I found Major Wilcox of the Canada Volunteers[189] in the same situation as myself; two or three miles further, I found Captain Morton of our regiment snug in a fine little bay.[190] I dined with him on some fine ducks that he had shot that morning. I arranged with him to have all my men who were able to march, join him as soon as the weather moderated, and march with his effective men to Sackett's Harbor; and to have him row down and take off all my disabled men and women.

After a dangerous and fatiguing row of twenty days from Fort George, we arrived at Sackett's Harbor. In a few days we all assembled as army and after a reorganization, we passed the Grenadier Island.[191] According to the original plan of the campaign, we were to attack Kingston and destroy the British fleet if possible. But General Armstrong, Secretary of War,[192] joined us and changed [Page 39] the plan. He directed General Wilkinson to leave Kingston in our rear and drop down the St. Lawrence to take Montreal with the aid of General Hampton, then at Chateauguay with four thousand men, who were to join us at the mouth of the Salmon River on the St. Lawrence, with additional boats and provisions for the whole army.

It was, I think, a great military error to leave the enemy's fleet and a large land force in our rear; but the orders of the Secretary were to this effect, and they were obeyed. While preparing to descend the St. Lawrence two schooners, by general order, took on board Surgeon McNear[193] and all the sick and disabled men, and sailed in a storm for Sackett's Harbor. The gale that night was tremendous, and the vessels were cast away on a reef or rocks, one mile from the main land. It was reported the next morning, and I volunteered (to General Boyd then commanding on the island)[194] to go to their rescue. He said that it was impossible in such a storm, but that if I would undertake it I might have as many men and boats as I pleased.

I took three Durham boats[195] and thirty of my own men, several of whom were old sailors. With great difficulty we reached the vessels and found them lying on the rocks with the sails blowing in every direction and the breakers making a full sweep over them. We boarded to leeward, but saw no one at first. I finally found the captain of the first vessel in an upper birth in the forecastle; I called and he answered, but refused to come up until I went down and beat him up with the flat of my sword. Dr. McNear was in the cabin in water up to his shoulders. We found everything afloat in the hold, including many dead men. They had got at the hospital stores, and were all drunk. We made thirteen trips to the shore and landed all, dead and alive, from both vessels, and unbent the sails and made a covering for them. In the afternoon, Dr. Ross[196] came from the island and took charge of them while I returned to Grenadier Island and reported. A general order was issued stating the disaster and that, the storm having abated, the dead and living were landed by boats from the garrison. It was now about the 1st of November, the men being badly supplied, their sufferings were great. There was clothing at Sackett's Harbor, but the Quartermaster had been unable to procure any. Our Majors, Huyck and Malcom, differed in most things. I consulted with them, and recommended an application in the name of the [Page 40] regiment to Colonel I. [sic] B. Preston[197] to take command of ours, he being without a regiment in consequence of having declined the command of a new one. We procured the sanction of the General, who issued complimentary orders to the Colonel and to the Regiment — to the latter for having offered, and to the former for having accepted the command. The Colonel asked me to go to Sackett's Harbor to try to procure some clothing. Taking a requisition, I started, and after a long row on a stormy night, I arrived at Sackett's Harbor, saw General Wilkinson on board of the "Lady of the Lake,"[198] and, after an altercation, succeeded in getting an order for all the clothing in store and started back with it to the island.

The dispute and explanation made Wilkinson and myself particular friends. As I am writing an account of my own experience, rather than a history of the war, it is unnecessary for me to excuse myself for speaking so often of my own movements. Everything being arranged, the army left Grenadier Island about the 5th of November in the same order in which we left Fort George, under convoy of some of Commodore Chauncey's squadron which proceeded with us as far as Frenchman's Creek, and then returned to Sackett's Harbor.

We passed through the Thousand Islands. They are so close to one another as to require skillful pilots; these we had, and notwithstanding sunken rocks, false currents, etc., passed safely through.

As soon as our convoy had returned to Sackett's Harbor, a British squadron of gunboats left Kingston and followed us down the river. Colonel M____ marched from Kingston[199] at the head of twenty-two hundred men, regulars and militia, and hung on our rear. We passed down the river to within ten miles of Ogdensburg,[200] where we landed on the American shore to avoid passing Prescott[201] in boats so full of men. The main body of our army marched by land, while a sufficient number of men and officers remained in the boats to take them down.

The senior Captains of the regiments were ordered to join Gen. Jacob Brown at two o'clock P. M.. We dropped down the river, noting the channel and the currents, until within round shot range of the guns of Prescott, when we returned to camp. Prescott was a fortification with a dry ditch and drawbridge with bombproof sufficient to cover five hundred men, as I ascertained by examination in 1820. It was a strong post, mounting twelve [Page 41] thirty-two pounders in barbette,[202] with an eight-gun water battery; it was worth taking. Our orders were to remain in camp until twelve o'clock, then, replenish our fires, drop down quietly, and pass Prescott with muffled oars. The night was a little hazy, but not dark.

My boat was ahead, and it was as large as a schooner. As soon as we were discovered, a heavy tire was opened upon us without effect, striking only one boat, killing one man, and wounding two. They kept up their tire until after we had all passed.

They fired, during the night, fourteen hundred rounds at our six hundred craft. We dropped down to White House,[203] where we ferried the army across to the Canada shore and swam the horses. We were now near the head of the Longue Saut rapids;[204] the enemy was in possession of the shore. Six strong detachments marched for a point five miles above Cornwall[205] at the foot of the rapids to secure the passage. When they arrived they were to notify us by the roar of cannon, but the first roar of cannon that we heard was from the enemy which had followed us from Kingston. They attacked and drove in our pickets and appeared in force.

Our line of battle was soon formed; and the battle of Williamsburg, or Chrysler's [sic] Field, began.[206] Quartermaster-General Swarthout, acting as Brigadier,[207] advanced his brigade from our right; the enemy fell back in line with their forces extending from the woods to the river.

The action became warm and general. Major____, who commanded the boat guard of five hundred men[208] with whom he had dropped down past Prescott, requested me to go to General Boyd on the field for orders. I reported to General Boyd; his answer was to remain and protect the boats, baggage, women, etc. I reported to Major Upham,[209] who dismissed the guards to their different bouts. Being senior Captain, my command was in the rear nearest to the enemy and the field of battle. In a few minutes, Col. N. Pinkney, aide to General Wilkinson,[210] who was at that time sick on board of his vessel, directed me to form my guard and march to the field. I immediately formed, eighty-six strong, including one of my pilots who volunteered (a fine fellow he was).

When we approached General Boyd near the field, he said, "Rush on my 'jolly snorters,' you are wanted."[211] Before forming into line, I halted a moment to let my men go into action coolly.

[Page 42] Giving them a few directions, I marched them into line. We met Colonel Cutting[212] with his regiment helter skelter, just broken out of line. "Colonel," said I, "where are you going?" Said he, "my men will not stand;" "but," said I, "you are leading them." They went off to the boats and I took the place of the regiment with my little detachment of eighty-six. I soon saw Major Malcom, and got permission to take position in a field on the enemy's right flank, where my "buck and ball" told well. My position was within two hundred yards of the right flank; my men kneeled on the left knee behind a stone wall about two feet high. I was already wounded by a musket ball passing through my left arm two inches below the socket. I received the wound while advancing, and, no doubt from excitement, believed it to be a chip that struck my arm. Both armies made several attempts to charge bayonets; but, as is almost always the case excepting when storming batteries, when one side charged the other fell back, and *vice versa*. At the beginning of the action, we had two pieces of field artillery under Lieutenant Smith on the river road on our left. They were taken by the enemy, and Smith and his men all killed.[213] I did not know that they were taken until the battle was over, although I saw a detachment of dragoons wheel up to them, fire their pistols, wheel again, and retreat on receiving a fire. I supposed it to be a British attack, but found, afterwards, it was an attempt to retake the pieces by Lieut.-Col. Worth;[214] but it was a complete failure. Our line was now out of ammunition for a short time, sixty-four rounds having been expended. The enemy formed in line and gave my little detachment a galling fire until again diverted in front. The ground was well contested for four hours on plain open ground, then the firing ceased on both sides.

There were fifteen hundred Americans against twenty-two hundred of the enemy. My wound being very painful on account of the pressure of my coat on the swelling arm, I gave the command of my sixty-three men — twenty-three having been killed — to my First Lieutenant Anderson.[215] Our troops were returning to the boats, the enemy having fallen back. Believing the field pieces before mentioned to be in the hands of our people, I walked towards them, intending to return to the boats by the road; I was very near when I discovered the British uniforms. I immediately turned and walked leisurely towards the retiring troops. They did not fire at me or pursue me until I rose from a ravine; they [Page 43] then fired, and one shot was returned by my servant, Williams, who was seeking me at some distance among the dead. He did not suspect that it was I, as he had heard that I was among the killed. It was with great difficulty, with the use of only one hand, that I got out of the ravine. I fell in with a man who had a horse and a keg of ammunition; I took the horse, and finally reached the boats. The horse was led part of the way by a camp woman. The loss in killed and wounded on both sides was not less than eight hundred, about equally divided. The result is not generally understood by our people, and it is as often called a defeat as a victory.

It was a complete victory.[216] We proceeded down the river, but Colonel Atkinson[217] came expressly to inform General Wilkinson that General Hampton could not meet us at Salmon Creek with men and supplies as agreed; for he had made an attempt to take or destroy a force at old Chateauguay, at the junction of that river and the St. Lawrence, been defeated, and fallen back on Plattsburgh.

This put an end to the campaign. We encamped at the junction of Salmon Creek and the St. Lawrence for some time. The place is now known as Fort Covington, named for General Covington who was killed at Chrysler's Field.[218]

I was invited to take up my quarters at the house of Dr. Mann.[219] I procured a horse which was led by my faithful servant Williams. When we arrived at Hitchcock's Tavern,[220] we met a small party at dinner, among them Miss Charlotte Bailey of Plattsburgh,[221] your mother, who was then visiting her uncle, Dr. Mann. I finally reached Dr. Mann's house, where I had a comfortable room, and was shown every attention by the family. General Wilkinson and several other officers stopped at the Doctor's house for a few days on their way to Malone.[222]

My wound had been neglected, and I had taken a severe cold by remaining on duty. A fever ensued, and I suffered everything but death. At one time, the Doctor feared that he could not save me. The army fell back on Plattsburgh early in December, and in March, I was so far recovered as to be able to follow.[223]

In this month I was married to Charlotte, daughter of Judge William Bailey.[224] I was then under orders to proceed as a witness before a Court-Martial sitting at New York for the trial of General Gaines.[225] When at New York my order was extended at my request, by Governor Tompkins, allowing me to go to Washington to settle my recruiting accounts. I returned to Plattsburgh in the [Page 44] beginning of April. In May, 1813 [sic: 1814],[226] the army advanced to Odletown, within three miles of the British outposts, near Champlain. At this time Commodore. McDonough advanced his squadron to Chazy.[227] There was some skirmishing with the enemy's outposts, but nothing of importance occurred during the two or three weeks that we remained there.

I was soon ordered to Plattsburgh. All the forces that could he spared were forwarded to the relief of General Brown,[228] who was closely besieged at Fort Erie.

We had a fatiguing march of seventeen days; then, we received news of the successful sortie at Fort Erie, and the victories of Chippeway [sic: Chippawa or Chippewa] and Lundy's Lane,[229] and it became unnecessary for me to go up the lake. We went into winter quarters and spent the winter of 1814 in tents. In the spring, General Brown told me that I was entitled to promotion, that he would send me to New York on special duty to return by way of Plattsburgh, at which place I was to take command of the companies of the 13th and march them to the regiments. On my arrival at Plattsburgh, I reported to General Macomb.[230] He said that he would add a company to the two of the 13th and give me the command as a battalion.

But he [Macomb] suggested that I might like to go to Franklin County for a time, as my wife was there visiting Dr. Mann, and I could procure valuable information there respecting the intended movements of the enemy in that direction, and in upper Canada. I went there and frequently rode to St. Regis and over the line, and I procured much valuable information at the risk of a halter for a neckcloth.[231]

I returned to Plattsburgh and reported to General Macomb. His battalion was not yet formed, and I remained there in camp until the news of peace was received. The men enlisted to serve eighteen months, and therefore, during the war, were to be discharged. But no money had been provided to pay the armies.

They were mutinous, and many of the companies refused to obey their officers. Among them, the two companies of the 13th claimed their pay and discharge, and when ordered to unstack arms, they disobeyed their officers to a man. I was called upon to address them. I reminded them of their services, and called on them not to tarnish the honor they had won by duty and discipline. I told them to continue to obey orders and to keep up the usual [Page 45] discipline for which the regiment had been noted, and that as soon as funds were received by the paymasters they should be paid, and such as were entitled to it should receive their honorable discharge. I now gave the word, "unstack arms." They complied, and I marched them over to the artillery who were under arms. We there drummed out of camp all who disobeyed the orders of their officers, and many of those who regretted their disobedience, were restored at my suggestion. In June, a board of officers was formed at Washington to reduce the army to a Peace Establishment,[232] and a rule was made that *none but effective officers should be retained.* All who had been wounded or disabled should be discharged with an allowance of three months gratuitous pay — reversing common sense and common justice rule that all who had been wounded or disabled should be retained to form skeleton regiments, which could at any time be filled with recruits.

Many of the officers petitioned to be retained, and some were retained and reduced one grade. Many of us met and determined, after debate, that as we had seen service and done our duty, we would neither petition to be retained, nor accept with reduced rank. At the request of the Colonel, I remained on duty with my regiment until September.[233] I then went to Washington to close my accounts. Paymaster-General Brant refused,[234] at first, to pay me beyond the fourteenth of June. After considerable difficulty, I obtained pay to the tenth day of September in Pittsburgh banknotes, which sold at eleven per cent [sic] discount at New York. So I lost eleven per cent. [sic] on eleven months' pay. Thus were we rewarded.

I now returned to private life at the age of thirty -eight. After having settled my accounts at Washington, I returned to Plattsburgh and procured board for my wife and child — my daughter Henrietta — at John Palmer[']s.[235] I then went to New York with about one hundred dollars to begin the world anew. After several weeks, and with much difficulty, I became engaged in the auction business.

I was aided by a friend[236] for whom I had laid the foundation of a fortune which he had realized during my absence. My prospects looked encouraging, when a former lieutenant of my regiment, Charles Mitchell,[237] called upon me for aid. He was destitute of means to pay even his board. I gave him what help I could, promising to pay his necessary expenses until he could procure a situation. He repaid me by robbing me and forging my name for three checks, two of which were on the Manhattan Bank. They **[Page 46]** could not determine whether the-checks were forgeries or not. I had great trouble, and nearly lost my reputation — always dearer to me than money.

Mitchell, to divert my attention from himself, wrote me threatening letters, saying that he had once been in my power and that I had tyrannized over him; that I was now in his power, and that he had robbed me, forged three of my checks, would forge more, would ruin me, and perhaps take my life. He forged checks of Dr. Silas Lord,[238] was detected, confessed all to me, and was sentenced by Judge Radcliffe,[239] a friend of his family, to leave the State and never return.

The discharged soldiers were continually calling on me for advice in respect to their claims for back pay, pensions, and bounty lands. Several agencies commenced the business, and I among the rest. I arranged to co-operate with Col. Joseph Watson[240] who opened an office at Washington, and we did a large business together in that way. I went on quite prosperously, and in 1817 took and furnished a small house in Walker street, went to Plattsburgh and brought down my wife. We lived there for some time, then removed to the corner of Canal and Mercer streets; we remained there two years, then removed to a three story house in Canal street where we lived until 1825, when I bought the house number 45 Mercer street and remained there until I bought Judge Vanderpool's [sic: Vanderpoel's] place at Kinderhook.[(241)] I was elected a member of the Assembly[(242)] at New York in 1828, re-elected in 1830 for '31, again for '32, '33, and '34 (in '29 I was not elected). I was once nominated assistant alderman, and once alderman of the eighth ward; I declined both nominations. The year after I removed to Kinderhook (1834) I was elected President of the village.[(243)]

During the time of my office, I received and addressed Martin Van Buren when Vice-President, and again on his return at the end of his term as President.[(244)]

I have now brought this sketch of the principal events of my long life to a period from which you are all acquainted . I have been as brief as possible, though sufficiently tedious to those not personally interested. I have omitted a great variety of incidents arising out of the events of the war, as not material in this sketch of my life.

I have no memoranda to aid my memory, and this accounts for the omission of particular dates.

[Page 47]
ADDITIONAL REMINISCENSES [sic]
OF THE
WAR OF 1812

In March, 1812, Lieutenant Valens [sic: Vail][(245)] accompanied me from New York to my recruiting rendezvous of the 5th district at Willsborough, in Essex County.

He was a fine young officer and amiable in disposition, but he knew nothing of military tactics. I appointed him my adjutant. He was desirous of acquiring a knowledge of his duty, but when speaking of it, he used to sigh and say, "It will be of no great service to me, for I have a strong presentiment that I shall be killed in the first engagement." I used to ridicule the superstition, and endeavor to convince him of the folly of entertaining it. But it so happened that Valens was attached to the second battalion of the 13th, and at the battle of Queenstown, when crossing from

Lewiston to the attack, he was heard to be in prayer in the bow of the boat; and it was not twenty minutes after landing that he was shot down by an Indian and scalped. George Helembold, my first sergeant, had the same presentiment. He was of the party that crossed the Niagara river below Black Rock to storm the British batteries.

At first, he supposed that he must be killed in accordance with his presentiment, but he soon found his error for he was brave and he did his duty manfully, and he was not killed, but taken prisoner. He soon returned on parole, with others, and being a printer, I allowed him to work in the printing office at Buffalo[246] until exchanged. At the time, the officers of the navy boasted of their exploit in taking the two British brigs from the protection of Fort Erie.[247] The movement was made under Lieutenant Elliot of the navy,[248] but the force was made up equally of soldiers and sailors. Helembold wrote and published a paragraph giving each corps equal credit for the part it took in the exploit, and signed it, "a young soldier." Midshipman Tatnall[249] called at the office to see the editor. He was absent and Helembold was in charge.

[Page 48] Helembold inquired the midshipman's business. Tatnall said, "I wish to know who is the author of a piece signed 'a young soldier.'"

"What is your object, sir?"

"Why, if he is a gentleman, I will treat him as such; and if not, I will treat him accordingly."

"Please state what, in your estimation, would entitle him to be considered a gentleman."

"Why, if he is a commissioned officer he is one, of course."

"But suppose, sir, that he is but a warrant officer like yourself."

"There are no warrant officers in the army."

"You are in error; all sergeants and corporals have warrants, and I have the honor to hold one as orderly sergeant of Captain Myers' company; and I fear, sir, if you were to know the name of the 'young soldier,' you could not treat him as a gentleman, and you must not meet him until he is promoted, and then, acting on your own principle, he will decide whether your rank will entitle you to his notice."

Tatnall went off in anger, and the matter ended.

When I joined the army at Greenbush,[250] in September, 1812, I found the officers too Democratic in their intercourse even for me, bred as I had been, in the Democratic school of Jefferson. They used to enter one another[']s tents, order the servants as though they were their own, and make free with the pens, ink, paper, and ——— drinks. In fact, it was "hail fellow well met." My idea of military etiquette and politeness revolted at such a mode of life. I stated my views to a number of the most gentlemanly officers, who accorded with me in opinion. We agreed to establish a suitable system of gentlemanly etiquette and immediately put it in operation. This added greatly to our comfort, but much to the disappointment of those who had supposed that all was to be in common. But in a short time, all were pleased with the change.

The day the army moved from Greenbush we merely crossed the river at Albany and encamped on the first night near the first turnpike gate on the road to Schenectady. During the night one of our soldiers stole from a countryman's yard an ox chain, and, at a halt on the following day, sold it. Before we reached Schenectady, one of his comrades stole it again. In fact, the chain was stolen and sold seven or eight times before the regiment reached Buffalo.

[Page 49] One night we chanced to encamp near a fine poultry yard. During the evening the men cooked their rations over their respective company fires.

One man carefully boiled something in a camp kettle, and a man from another company, seeing feathers on the ground, suspected that fowl were cooking, and procuring a kettle from his own company he hung it over the same fire. When the proprietor of the fowl was out of the way for a moment the man exchanged the places of the kettles, and when requested to take his rations to his own camp fire, he went off with great docility, taking the kettle filled with the fowl and leaving his own filled with water.

One night, while encamping near Utica, some of the officers applied to Colonel Christie for permission to go to the town, but their request was refused. In revenge, they persuaded all the soldiers, officers, and men, to shave their mustaches, to the great annoyance of Colonel Christie, who had particularly desired that they should be worn.[(251)]

When in command of the 5th recruiting district of Plattsburgh in July, 1812, a large body of militia came in. Two of the officers had a difference; a challenge was given and accepted, and the parties were to fight in the lake to avoid the law which disfranchised those who sent or accepted challenges in the State. The parties met according to agreement; but their friends had ingeniously arranged to load with blood instead of ball. At the given word, they tired; one of them fell, and was taken out of the water by his friend covered with blood. It was some time before he could be convinced that he was not mortally wounded.

One night while encamped at Flint Hill, three miles from Buffalo, I had just lain down in my tent on some straw covered with a buffalo skin, the whole resting on muddy ground — for it was in October, 1812, and everything was wet on and about us, as it had been raining more or less for ten days — I was called to meet the captains and subalterns of the brigade on the right of the line.

Captain Brooke (now General Brooke)[(252)] addressed the unlawful assemblage nearly as follows: "Gentlemen and fellow soldiers; we all came here to fight and to conquer the enemy. not to remain here suffering every want and privation, under the orders of General Smyth, useless to ourselves and to our Country. Six [Page 50] weeks I have been on the lines. My friends expected to hear from me before this; what can they hear, but that we are living inactively in our tents at Flint Hill in rain, mud, and want; eating grass seed and beef without salt, wearing soiled shirts for want of soap to wash them, and without candle light.["] Here, some one observed, "Brooke, you have no fortitude." "Have I not," said Brooke, .and putting the forefinger of his right hand into the blaze of the company fire before us, he held it there until it was deeply burned.

"Now, gentlemen," said he, "I propose that each captain take full command of his company, and that the senior captain of the brigade lead us to, and across the river, that we may take Fort Erie by storm, proceed to Chippeway, and thence to Fort George and take that-— and leave General Smyth to sleep the night out in his snug quarters."

All, for different reasons, agreed to this. The paymaster of the 5th said, "Gentlemen, General Parker,[253] my relative, is in camp, and I have no doubt that, if a deputation calls upon him, he will join us and lead us on to glory."

Accordingly, Brooke, three others and I, were appointed. We moved towards his marquee. But a private message was sent to General Parker to inform him of what was going on. The General received us very kindly, approved of the plan, and ordered refreshments. The effect of good, old brandy was to make the gentlemen very drowsy. They drank freely, and soon the whole mutiny and expedition were forgotten; and General Parker, an excellent man, never mentioned it, I believe, and it passed off harmlessly.

Once, while we were at Fort George, Major____ (since promoted)[254] was ordered to march with three companies of infantry to ascertain when the British picket guard, opposite to our picket on the swamp road, was stationed. Mine was one of the companies. We marched about a mile and halted. The Major told me to march on until I drew out the enemy's guard, and then to fall back on his other two companies.

I did so, and half a mile farther on I found the guard. It turned out in full force and I made a false retreat, but when I got back to my supplies there were no two companies, nor even a man to inform me where they had gone. I marched on and found them in camp. The Major said that he had supposed that I had returned by a different road, not recollecting that to have done so, I should have disobeyed orders. It passed quietly.

[Page 51] One night we encamped in a swamp. Lieut. J. C . Eldridge and I mounted a barrel of whiskey back to back for a lodging. In about an hour I heard a noise, and, looking around, I found that the men were pulling Eldridge out of the fire. It appeared that when we were both asleep, I overbalanced him and he fell into the fire. Fortunately, he was not much burned.

When encamped at Cumberland Head, Lake Champlain, I was walking on the bank one morning and I saw a soldier lying in a fire on the beach. I jumped down about eight feet and pulled him out — a little burned. He was not entirely sober, but he thanked me and walked off. Some years after he came to my office in Wall Street, New York, related the circumstance and said that I had, no doubt, saved his life. It is a great happiness to have it in our power to save a life.

I had a soldier named Clark[255] who was a noted thief. While encamped at Northfields, near Champlain, Clinton County,[256] he robbed the sutler of the 14th Regiment of all his goods, valued at fourteen hundred dollars. He gave away most of the goods. He was detected, tried, and condemned to have his pay stopped to make good what he had taken; and when the Paymaster was pay-ing the men, he would come to me with tears in his eyes and beg me for a dollar. I always gave it to him. He was an excellent sol-dier in all other respects, but he was by nature a thief.

Poor, weak human nature! We require discipline and firmness to resist our natural propensities.

It would naturally be supposed that, if any class of men is free from superstition, it is sailors and soldiers; but, as a rule, I have found the latter to be much under the influence of it. I had in my company a young Irishman whose bravery was never doubted; but while I was commanding the Cantonment at Williamsville, in the winter of 1812, a report was one morning current that O'Bryan had seen the Devil,[256] when on post at the sally-port leading to the graveyard. Fearing that others might feel alarmed on going on the same post, I sent for O'Bryan and said to him, "It is said that you saw the Devil last night when on post." He turned very pale and said, "I do not know whether it was the Devil, but something black, about the size of a horse without a head, came rolling towards me. I challenged, but received no answer; it was light, .and I plainly saw it still approaching; I sighted my piece to fire, but lost my strength and [Page 52] fell back against the Can-tonment where the relief soon after found me." I said, "Are you afraid to go on the same post, at the same hour tonight with me? Perhaps we may find that there was no Devil, but your own fear of spirits." He said, "Sir, I have never known fear, and I am will-ing to go with or without you — but I should prefer to have you go with me." I went, and we remained on post two hours; but the black devil in the shape of a horse without a head did not renew his visit, and my having been on post with O'Bryan was known to all the command, and it satisfied all that they might go on that post without meeting "His Satanic Majesty."

At the formation of the American Army of 1812, there was a great rush for appointments.

Many desired to join it from high minded notions of national honor, or patriotism; others from motives of pride and love of military show and splendor, and many for employment and pay. Many were recommended by Members of Congress without much regard to their qualifications for command; so that at the recruiting service, and the assemblage of the troops at Greenbush, in the autumn of 1812, we found a gathering of heterogeneous characters bearing commissions — men taken from every grade of society, from the highest to the lowest. But all felt the same pride. The most ignorant were the most jealous of their honor — so far, at least, as to take offence at the least circumstance or even word that seemed to them to reflect on their courage. Challenges to single combat were of frequent occurrance [sic]. There are, occasionally, insults offered, or injuries inflicted for which the law affords no satisfaction; and it then becomes necessary for a man to take the matter into his own hands. But in most cases, ducls are fought for trifling wrongs, or imaginary offences, and promoted by injudicious friends of the parties, from false pride and a desire to become conspicuous in such matters; when in ten cases out of twelve the affairs might be amicably adjusted by well disposed friends, and so save the lives and the honor of well mean ing but hot-headed men. It was my lot to be engaged in several affairs of the kind both before and after I became an officer in the regular army, and I look back with great satisfaction at many differences amicably arranged between those who, without a friend so disposed, would have lost their lives in silly combat, or have taken the-lives of their opponents.

[Page 53] But in some few cases of aggravated controversies, I did not succeed in making settlements, and duels resulted. (See appendix.)

In a few cases challenges were refused, and the person refusing was justified by the whole army.

When the army lay at Fort George, Upper Canada, in 1813, Lieut.-Col. Winfield Scott was ordered out with a command of men to the head of Lake Ontario to take or destroy the British public stores at Burlington Bay — in which enterprise he was particularly successful. Colonel Homer Milton[258] considered that as he ranked Scott the command belonged to him. A correspondence took place which resulted in a challenge from Colonel Milton. Scott declined in words to this effect:

"I have no personal difference with Colonel Milton; if a command has been given to me in preference to him, he should apply to the commanding general who directed his adjutant-general to give me the order. But if the Colonel desires to test my courage, I prefer to show it to him in the face of the enemy when we are at the head of our respective commands."

When the troops under Gen. Stephen Van Rensselaer crossed the Niagara to take Queenstown, a lieutenant of light artillery named Randolph,[259] who had no command in the expedition, jumped into the boat of one of the captains of the 13th Infantry and crossed over to Queenstown where he had command of the advance of the 13th Infantry.

The circumstance was mentioned to Lieut.-Col. Christie, who observed that he had no knowledge of Mr. Randolph, or that he had any command in the 13th Infantry.

Randolph, a fine, spirited young man sent a challenge to Colonel Christie, who was then at the head of his regiment near Fort Niagara preparing for the invasion of Canada.

Christie sent for a few of his captains to consult. We advised him — considering their relative ranks, the circumstances giving rise to the challenge and the responsibility of his position at that time — to decline. He was also posted, but he was justified by most of the officers.

Randolph, having spoken of the matter, was informed that there were eight captains in Christie's regiment who resented the same remarks that their Colonel did; that all were nearer his rank, and that h might call on any of them, if he pleased. No more was said or done.

[Page 54]
APPENDIX.

___

One of the duels referred to by Major Myers was between Surgeon Bronaugh[260] and Major Stonard;[261] copies of the letter of Dr. Bronaugh conveying the challenge is given below, and also a copy of the arrangements made by the seconds for the hostile meeting.

___

CAMP, NEAR FORT NIAGARA,
12th May 1813

SIR: Your contemptible evasion of a direct challenge I shall take the liberty of construing into an act of base cowardice unbecoming the station you hold in the army, and an aggravation of the offence for which I have demanded an honorable reparation at your hands. The whole history of the affair of which I complain characterizes you a victim of a feebleness of judgment and a vindictiveness of disposition that fitted you to be the tool of a designing scoundrel, more base and cowardly perhaps than yourself, and in which he manifested himself superior to you only in consummate hypocrisy. It is now obvious to every unprejudiced person that Colonel Coles [sic], your co-adjutor in the holy work of "motives of duty and the good of the public service," was the prime mover of the charges preferred against me; that he was anxious, in conjunction with you, to get the officers of the regiment to sign them, but they, feeling too much independence to sacrifice their judgment at the shrine of malice, refused to unite in a coalition, the object of which was my downfall, and not "motives of duty and the good of the public service." And no one being found supple enough to put his name to those charges but you sir, and you I hold responsible for the outrage committed on my feelings — your devotion to Colonel Cole [sic] blinded you to a sense of justice, and your milk and water apology (?) made through my friend convinces me that you have adopted the maxim "that the holiness of the end will sanctify the most dishonorable means," and that you hoped for absolution from guilt if by serving the purposes of one-man you could injure another whom you had professed to respect.

[Page 55] I am not to be appeased, sir, by the atonement you have attempted to make, it is not equivalent to the-injury, and unless you give the satisfaction I demand I shall denounce you to the world as a base, rotten-hearted coward, who is neither soldier, officer nor gentleman, and I shall make my first letter addressed to you and this also, through the medium of the Buffalo Gazette, a matter of publicity to the world.

Yours, etc.,
JAMES C. BRONAUGH,
Surgeon, 12th Infantry.

Major JOHN STONARD,
20th Infantry.

———

13th May, 1813.

The parties will meet tomorrow morning at 7 o'clock on the lake shore about one quarter of a mile from camp, they will take their distance at ten paces, face to face, they will stand with their pistols pointed downwards; the question is to be asked "are you ready?" on an answer in the affirmative the word "fire" is to be given; they are to raise their pistols and discharge them; it is to be considered that they are to discharge their pistols within one minute of the word "fire" or lose their fire. After the first shot they are to advance one pace each, after the second one more pace each, and so on until one of the shots takes effect, and so on until one of the parties declares he is unable to fire; each of the parties to be accompanied with two friends and a surgeon and none others.

R. WHARTENBY,[262]
M. MYERS.

———

With the above remarks upon duelling the series of letters from which these reminiscences are selected, closes; and with the end of the war, Major Myers' military career ended. He afterward resided in New York City and Schenectady; he twice represented New York City in the Legislature of the State for six years, and served two terms as Mayor of Schenectady;[263] he was a candidate for Congress when eighty-four years old.[264] He was present at the inauguration of Washington as President by Chancellor Livingston, and retains a vivid recollection of that event. He was probably one of the most prominent Masons of his day, having been Grand Master and Grand High Priest, respectively, of the Grand Lodge and Grand Chapter of the State of New York.[265]

[Page 56] Major Myers died at Schenectady, January 20, 1871. The *Schenectady Times*,[266] on the occasion of his funeral, says: "Major Myers was in many respects a remarkable man. He was possessed of a clear mind, strong will, and the fact that, with all the hardships incident to the life of a soldier in war of 1812, he lived to be nearly ninety-six years of age, is proof that he possessed a strong and robust constitution. His physical appearance was striking. No stranger ever met him or passed him on the street without noticing his appearance; he was of very large proportions and had .a clear and keen black eye, giving evidence of the strong intellectual power of the man. As Mayor of this city he added dignity to the office, and brought all the power of his common sense and an indomitable will to bear against wrong, and in favor of right and justice."

Errata.

Page 36, third paragraph, and fourth line, Colonel Butler should read Colonel Boerstler,.

Page 37, fifth line, Butler should read Boerstler.

Notes on Part Two:
The Reminiscences of
Mordecai Myers, 1780-1814;
Captain, *13th United States Infantry,*
1812-1815

(1) Theodorus Bailey Myers (1821-1887), Mordecai Myers's eldest surviving son, was an attorney, philanthropist, historian, and a fixture in New York City political and social circles. Catalina Mason Myers Julian James, *Biographical Sketches of the Bailey-Myers-Mason Families, 1776-1905* (Washington, D.C., 1908), pp.23-36.

(2) Myers's parents were Benjamin Myers (1733-1776) and Rachel Myers (1745-1801). "Items Relating to the Jews of Newport," *Publications of the American Jewish Historical Society,* (1920), Vol. XXVII, p.198. Hereafter cited as PAJHS; David De Sola Pool, *Portraits Etched in Stone – Early Jewish Settlers, 1682-1831* (New York: Columbia University Press, 1952), p. 285.

(3) Hellevoetsluis was once the home port of the Dutch fleet. Packet boats had been carrying passengers and letters across the North Sea between Hellevoetsluis and Harwich, England since the 14th century. Like many travelers of the period, the Myers may have proceeded from Harwich to Falmouth and on to America. "Hellevoetsluis," in Van der Ar, *Aardrykundig Woondenbock van Nederland* (1884), pp. 388-95. Photocopy courtesy of Dr. W. Chr. Preterse, Director of the Municipal Archives, Gemeentelijke Archiefdienst, Amsterdam, Holland. Translated for the editor by Alexander Lenten.

(4) Newport, Rhode Island had a well-established Jewish community since 1739. See Morris A. Gutstein, *The Story of the Jews of Newport: Two And A Half Centuries of Judaism, 1658-1908.* (New York: Bloch Publishing Co., 1936), et passim; and Sheila Skemp, *A Social and Cultural History of Newport, Rhode Island, 1720-1765* (Ann Arbor, Michigan: Xerox University Microfilms, 1974), et passim.

(5) Myers birthplace was in a house located on Marlborough Street. St. Paul's Methodist Church has occupied the site since 1806. PAJHS, XXVII, p.14; Malcolm Stern, "Myer Benjamin and His Descendants: A Study in

Biographical Method", *Rhode Island Jewish Historical Notes* (1968), Vol. 5, No.2, p. 143, n.18. Hereafter cited as "Stern."

(6) This is something of an exaggeration. Benjamin Myers spoke Hungarian, German, Yiddish, had a working knowledge of Hebrew, and learned English after settling in American in 1757-58. Stern, 134-35, 137, n.6; S. Broches, *Jews in New England: Six Historical Monographs* (New York: Bloch Publishing Co., 1942), p. 381; Jacob Rader Marcus, *The Colonial American Jew* (Detroit, Michigan: Wayne State University Press, 1970), Vol. II, pp. 1062-63.

(7) Ezra Stiles (1727-1795) was the pastor of Newport's Second Congregationalist Church from 1755 to late 1776. He befriended many of the members of Newport's Jewish community and studied the Hebrew language and Jewish ritual. Franklin Bowditch Dexter, ed. *The Literary Diary of Ezra Stiles, D.D., LL.D.,* 3 volumes (New York: Charles Scribner's Sons, 1901), et passim; George Alexander Kohut, *Ezra Stiles and the Jews: Selected Passages From His Literary Diary Concerning Jews and Judaism With Critical and Explanatory Notes.* (New York: Philip Lowen, 1902). Also see "Ezra Stiles", *Encyclopedia Judaica* (Jerusalem, Israel: Keter Publishing House Jerusalem Ltd., 1972), XV, p. 403.

(8) Benjamin Myers died on Wednesday evening, November 20, 1776, at age 43 from complications arising from his having broken his arm several weeks earlier. Stern, p. 135, n. 23; pp. 138-39; PAJHS, p. 198, p. 350.

(9) Myers means "occupied", not "evacuated." The British occupied New York City on September 15, 1776. They evacuated Newport on October 26, 1779, after which Rachel Myers took her family to New York City. Jacob Rader Marcus, *American Jewry – Documents – Eighteenth Century* (Cincinnati, OH: The Hebrew Union College Press, 1959), p. 274.

(10) The Myers' stay in Nova Scotia/New Brunswick from 1783 to 1787 was anything but pleasant. The family struggled unsuccessfully to obtain good land and subsisted on government provisions until returning to America in 1787. Neil B. Yetwin, "American-Jewish Loyalists: The Myers Family of New Brunswick", *The Loyalist Gazette* (Toronto: The United Empire Loyalists Association of Canada), Vol. XXVII, No. 2, Fall, 1990, pp. 13-18.

(11) Richard Lippincott (1745-1826) fought with the Loyalist Provincial Corps during the Revolution. Joshua Huddy, who fought in the Continental army, was implicated in the murder of a Loyalist named Philip White, one of Huddy's relatives. Huddy was captured and brought to New York City, where the Board of Associated Loyalists authorized Lippincott to execute him in retaliation for White's murder. Lippincott hanged Huddy at

Sandy Hook, New Jersey on April 12, 1783, which attracted attention in American and Europe. He was court-martialed but acquitted on the grounds that he had acted no from "malice or ill will, but proceeded from a conviction that it was his duty to obey the orders of the Board of Directors of Associated Loyalists." Egerton Ryerson, *The Loyalists of America and Their Times from 1620 to 1816* (New York: Haskell House Publishers Ltd., 1970). Rpt. of 1880 ed., II, p. 194.

(12) Benedict Arnold (1741-1801) was the American General who defected to the British side and was involved in an unsuccessful plot to hand West Point over to the British. Arnold and his son emigrated to St. John, New Brunswick (partitioned from Nova Scotia in 1784) where he acquired a town lot and established a store in November, 1785. In May 1786, Arnold went on a trading voyage to England and the West Indies, leaving the store to be run by his partner, Monson (or "Munson") Hoyt (or" Hoit", "Hait"), who had served with the Prince of Wales Regiment during the Revolution. Arnold had taken out a 6000-pound insurance policy on the store and stock and returned to St. John in July. 1787. When a fire destroyed the building, Hoyt accused Arnold of arson and sued him. Arnold initiated a counter-suit for libel, but was awarded only 20 shillings. He left New Brunswick and Hoyt returned to the United States. *Dictionary of American Biography* (New York: Scribner, 1958-1964), vol. I, 362-67. Hereafter cited as *DAB;* Sharon Dubeau, *New Brunswick Loyalists: A Bicentennial Tribute* (Agincourt, Ontario: Generation Press, 1983), p. 74; North Callahan, *Flight From the Republic: The Tories of the American Revolution* (New York: The Bobbs-Merrill Company, Inc., 1967), p. 64.

(13) John Paul Jones (1747-1792) was America's first well-known naval fighter in the Revolution, having responded, "I have not yet begun to fight" to the suggestion by a British captain that he surrender the *Bon Homme Richard* to *H.M.S. Serapis.* Jones had left Paris in spring, 1787, for Copenhagen, but while in Brussels decided to return to America during the summer and fall of 1787 when Congress gave him a gold medal for his services. Jones was in Hartford, Connecticut, in 1787 and he may have traveled from there to New Haven to board a ship. Thirteen-year-old Mordecai likely encountered him after September 17, 1787. *DAB,* V, 183-88; *Connecticut Courant,* September 17, 1787: Issue 1182, p. 3; *Middlesex Gazette,* September 24, 1787, Vol. II, Issue 99, p. 2; *Litchfield Monitor/The Weekly Monitor,* September 24, 1787, Issue 144, p. 2. Photocopies courtesy of the Connecticut Historical Society.

(14) Robert Richard Randall (1740-1801) was a wealthy merchant, land-owner and U.S. Vice-Consul to China. Randall bequeathed 21 acres, stocks and other properties to be used for a "Marine Hospital" to be called "Sailors'

Page 133

Mordecai Myers, U. S. Army. War of 1812

Snug Harbor", where 950 retired seamen found a home from 1833 to the 1890's. *DAB*, VIII, pp. 348-49; Ira K. Morris, *Memorial History of Staten Island* (New York: Memorial Publishing Company, 1898-1900), Vol. II, pp. 413-22; *New York Times*, August 14, 1912, SM14.

(15) William Cunningham was Provost-Marshall of New York during the Revolution. By 1778 he was in charge of American prisoners in New York and Philadelphia. 2000 of the prisoners under his charge starved to death (he had sold their rations) and more than 250 were hanged without trial. There were unsubstantiated reports that Cunningham was later executed in England for forgery after confessing his crimes. Daniel Curry, *A Historical Sketch of the Rise and Progress of the Metropolitan City of America.* (New York: Carlton & Phillips, 1853), pp. 137-42; Danske Dandridge, *American Prisoners of the Revolution* (Charlottesville, VA: The Michie Company, 1911), pp. 33-37.

(16) Myers meant James "Commodore" Nicholson (1737-1804), a leader of the anti-Federalists in New York City. Nicholson was an associate of the Clintons and Livingstons, one-time U.S. Secretary of the Treasury, and father-in-law of the longest-serving U.S. Secretary of the Treasury Albert Gallatin (1761-1849).

(17) Thomas Jefferson (1743-1826) was the principle author of the Declaration of Independence, third President of the United States, and founder of the University of Virginia. Jefferson worked closely with Aaron Burr and tied with him for first place in the Electoral College, leaving the House of Representatives to decide the election. The Federalists, led by Alexander Hamilton, still had some power, and since Jefferson was perceived by Hamilton as less of a political evil than Burr, Jefferson became President and Burr Vice-President.

(18) Alexander Hamilton (1755 or 1757-1804) was the first Secretary of the U.S. Treasury. He was killed by Aaron Burr in a duel on July 11, 1804.

(19) Timothy Pickering (1745-1829), a Federalist, was the third U.S. Secretary of State under both George Washington and John Adams in 1795-1800. *DAB*, VII, pp. 565-68.

(20) Oliver Wolcott, Jr. (1760-1833) was a federal judge appointed by George Washington to succeed Alexander Hamilton as U.S. Secretary of the Treasury during 1795-1800. Wolcott later served as the governor of Connecticut. *DAB*, X, pp. 443-45.

(21) In 1785, Thomas Greenleaf (1755-1798) became the managing editor of the *New-York Journal and Patriotic Register,* an important anti-Federalist newspaper, and purchased it in 1787. On July 26, 1788 a Federalist mob

celebrating New York's ratification of the Constitution earlier that day broke into Greenleaf's shop and destroyed much of his type. Greenleaf was the printer for the New York State Senate Journal for a time, and was once charged with using words to incite a local mob to riot over moving the nation's capital out of New York City. He also published *The Argus or Greenleaf's New Daily Advertiser*. Greenleaf died on September 20, 1798 as a result of the yellow fever epidemic that swept through New York City that year. James Edward Greenleaf, *Genealogy of the Greenleaf Family* (Boston, 1896), p. 65, p. 68.

(22) Myers is probably confusing this with Newburgh, NY, George Washington's one-time headquarters, or it was a misreading of the "g" as a "y". In any case, Myers meant Poughkeepsie, NY, the second capital of New York State after the Revolution. In 1788, Alexander Hamilton, John Jay and George Clinton held debates and helped to ratify the U.S. Constitution at the courthouse on Poughkeepsie's Market Street, making New York the 11th state in the new union.

(23) Robert R. Livingston (1746-1813) was for 15 years New York State's Chancellor , the highest judicial office in the state. As U.S. Minister to France, he negotiated the Louisiana Purchase and was the first Grand Master of the Grand Lodge of the State of New York. Livingston administered the oath of office to George Washington on April 30, 1789.

(24) Myers recalled the New York City procession celebrating the ratification of the Constitution as having occurred in 1799, but it took place on July 23, 1788. The procession included 5000 men and boys representing more than 60 professions and trades, and measured about one and one-half miles in length. Edwin G. Burrows and Mike Wallace, *Gotham: A History of New York City to 1898* (Cary, North Carolina: Oxford University Press, 1998), pp. 288-98; Martha J. Lamb and Mrs. Burton Harrison, *History of the City of New York: Its Origin, Rise, and Progress* (New York: A.S. Barnes and Company, 1896), Vol. II, pp. 525-26.

(25) Sebastian Bowman (or Baumann, 1739-1803) was a German-born military engineer who had served in the British army but switched to the American side after his marriage to an American woman. He helped organize the Sons of Liberty and was the last officer to leave New York City when the Americans evacuated on September 15, 1776. Bowman commanded the artillery at West Point, participated in the siege of Yorktown, and commanded artillery when the American army returned to New York City on November 23, 1783. After the war he became a grocer and wine merchant and was appointed Postmaster of New York by George Washington for his faithful service. Eliza Susan Quincy, *Memoir of the Life of Eliza*

S.M. Quincy (Boston, Massachusetts: John Wilson and Son, 1861), pp. 43-44; William Leete Stone (trans.), *Letters of Brunswick and Hessian Officers During the American Revolution* (Albany, N.Y. : Joel Munsell's Sons, Publishers, 1891), pp. 133-34; Frederic Gregory Mather, *The Refugees of 1776 from Long Island To Connecticut* (Albany,N.Y.: J.B.Lyon, Printers, 1913), 666n.

(26) Writing this in 1853, Myers is alluding to the debates over the slavery issue which culminated in the Kansas-Nebraska Act of 1854, which divided the nation and put it on the path toward civil war.

(27) Myers, then nearly 16 years of age, left New York City for Richmond in 1791-92 with his older brothers Benjamin and Abraham. Richmond, like Newport, Rhode Island, had an established Jewish community that made up about 1/6 of the population of 2000. Herbert T. Ezekiel and Gaston Lichtenstein, *The History of the Jews of Richmond From 1769 to 1917* (Richmond, Virginia: 1917), p. 35.

(28) George Wythe (1726-1806) was a signer of the Declaration of Independence, America's first law professor (at the College of William and Mary), and taught law to Thomas Jefferson, James Monroe, John Marshall, Henry Clay and others. Wythe held slaves but became an abolitionist. His grand-nephew, George Wythe Sweeney, tried to poison Wythe and two slaves he had included in his will. Wythe died but not before changing his will, leaving Sweeney nothing. Sweeney was acquitted because of a law Wythe had written which forbade the testimony of black witnesses. Sweeney later served time for horse stealing and disappeared. *DAB*, X, 586-89; Imogene E. Brown, *American Aristides: A Biography of George Wythe* (Rutherford, NJ: Fairleigh Dickinson University Press, 1981), et passim; Bruce Chadwick, "The Mysterious Death of George Wythe," *American History*, February 2009, pp. 36-41.

(29) Myers is referring to the January, 1793 election held in New York to elect 10 New Yorkers to the U.S. House of Representatives. In 1792, Congress re-apportioned the number of seats and New York's representation increased from 6 to 10. As a result, on December 18, 1792, congressional districts in the state were re-apportioned. The New York State legislature divided the existing 6 Congressional districts into 10 districts, and the elections were then held in June 1793.

(30) Four bills were passed in 1798 by the Federalists in the 5th U.S. Congress during America's undeclared "Quasi-War" with France, and signed into law by President John Adams: the Naturalization Act, Alien Act, Alien Enemies Act, and Sedition Act, all used to quash Republican editors, politicians and private citizens who criticized the Adams administration.

(31) Henry Brockholst Livingston (1757-1823) was a veteran officer of the American Revolution, attorney, Justice of the New York State Supreme Court, and Associate Justice of the U.S. Supreme Court. *DAB*, VI, pp. 312-13.

(32) Morgan Lewis (1754-1844) was an American Revolutionary officer who served as New York State Attorney-General, Judge of the Court of Common Pleas and of the Superior Court, Chief Justice of the New York State Supreme court and Governor of New York. Lewis served as Quartermaster-General of the Northern Army during the War of 1812, and as Major General commanded the defenses of New York City. *DAB*, VI, 222-23; David S. and Jeanne T. Heidler, eds., *Encyclopedia of the War of 1812* (Santa Barbara, California: ABC-CLIO, 1997), pp. 299-300.

(33) Aaron Burr (1756-1836) served as an officer in the American Revolution and went on to have a successful political career in the New York State Senate and Assembly, as Attorney-General of New York, United States Senator, and Vice-President under Thomas Jefferson. Burr killed Alexander Hamilton in a duel on July 11, 1804.

(34) A Democratic-Republican, George Clinton was a soldier and politician who was the first Governor of New York State (re-elected 5 times, 1777-1795), New York State Assemblyman, and 4th Vice-President of the United States, serving under both Thomas Jefferson and James Madison. *DAB*, II, pp. 226-28.

(35) Horatio Gates (1727-1806) was a retired British soldier who served as an American general during the Revolution. He took credit for the victory at Saratoga but was blamed for the defeat at Camden. After the war, Gates and his wife retired to an estate in northern Manhattan Island but remained active in New York society. Gates supported Thomas Jefferson's bid for the Presidency and served a single term (1800) in the New York State Legislature. *DAB*, IV, pp. 184-86.

(36) Marinus Willett (1740-1830) was a veteran of both the French and Indian War and the American Revolution. A cabinetmaker by profession, Willett was also known as something of a street brawler and became a leader of the Sons of Liberty, breaking into the New York City arsenal and seizing weapons after news of the Lexington and Concord reached New York City. After the war, Willett became an anti-Federalist associate of George Clinton and served as a State Assemblyman, Sheriff of New York City, election inspector, Superintendent of Construction of Fortifications of New York City, and Mayor of New York City after DeWitt Clinton was removed from office. 10,000 mourners attended his funeral. *DAB*, X, 244-45; Larry Lowenthal, *Marinus Willett: Defender of the Northern Frontier* (Fleischmanns, NY: Purple Mountain Press, 2000), et passim.

(37) Henry Rutgers (1845-1830) came from one of New York's wealthiest leading colonial families. A supporter of the Sons of Liberty, Rutgers was a member of the New York State Assembly, Vice-President and President of the New York Democratic Society and Presidential Elector. In his later years he raised money for the construction of the first Great Wigwam of Tammany Hall, acted as Regent of the State of New York, and a trustee of Princeton and Queens College, the latter re-named Rutgers University in his honor. *DAB,* VIII, 255-56; David J. Fowler, "Benevolent Patriot: the Life and Times of Henry Rutgers, 1745-1830" (Special Collections and University Archives, Rutgers, The State University of New Jersey, 2010), pp. 1-29.

(38) Ezekiel Robbins (1732-1816), a supporter of Aaron Burr, was a New York State Assemblyman and New York City Health Commissioner. Robbins was one of a group of Democratic-Republicans who were proposed for the State Assembly in 1800, including George Clinton, Horatio Gates, Henry Rutgers, Henry Brockholst Livingston, John Swartwout and others. James Parton, *Life and Times of Aaron Burr* (New York: Mason Brothers, 1858), 247-48; Matthew Davis, *Memoirs of Aaron Burr, with Miscellaneous Selections From His Correspondence* (New York: Harper & Brothers, 1836), Vol. II, pp. 251-52.

(39) Tammany Hall (or The Society of Tammany, Sons of St. Tammany, The Columbian Order) was a New York political organization founded in 1786 and incorporated May 12, 1789 as the "Tammany Society". Named after Tammanend, a Native American leader of the Lenape tribe, it was the Democratic Party's political machine that controlled New York City politics and helped immigrants (mainly the Irish) rise in American politics from the 1790's to the 1960's. By the mid-19th century it had been corrupted by graft. For a history of Tammany Hall see Gustavus Myers, *The History of Tammany Hall* (New York: Dover Publications, Inc. 1971), and Jerome Mushkat, *Tammany: The Evolution of a Political Machine 1789-1865* (Syracuse, New York: Syracuse University Press, 1971).

(40) Myers means Melancton (or Melancthon) Smith (1744-1798), a prominent member of the New York Democratic Society in 1793. Smith was a New York City delegate to the Continental Congress as well as to the first New York Provincial Congress. As a major in a company of New York militia known as the "Dutchess County Rangers" (during which he simulta-neously served as Sheriff of Dutchess County), Smith was instrumental in detecting, arresting and informing upon Loyalists and purchasing their forfeited estates. Smith moved to New York City in 1785 where he became a wealthy merchant, helped establish the Manumission Society, and was elected to the Continental Congress and the New York State Assembly. He became known as the "Patrick Henry" of the Constitutional ratifying

convention because of his "plain yet compelling oratory." Smith died during the yellow fever epidemic that swept New York City in 1798. *DAB*, XVII, 319-20; Nancy Isenberg, *Fallen Founder: The Life of Aaron Burr* (New York: Viking Penguin, 2007), p. 87, p. 126, p. 140.

(41) Major James Fairlee (1757 or 1759 – 1830), an intimate friend of George Washington, fought in the Revolution and settled in Albany, New York where he acquired land, traded in military bounties and bounty rights, and became an alderman. He moved to New York City in 1800, where he became a clerk of the Circuit Court of New York City for 30 years and was a founder of the Society of the Cincinnati. James Grant Wilson, ed., *The Memorial History of the City of New York, From Its First Settlement To the Year 1892.* (New York: New York History Company, 1893), Vol. III, p. 118, n. 1.

(42) Attorney Sylvanus (or Silvanus) Miller (1772-1861) was a graduate of Columbia College, New York County Surrogate, Public Administrator for the Common Council of New York, member of the New York State Assembly, and for 20 years a judge of the New York State Supreme Court. He is best known for having processed Thomas Paine's will on July 12, 1809. *Littell's Living Age. The Fifteenth Quarterly Volume of the Third Series.* Volume 71, p. 242, October, November, December, 1861 (Boston: Littell, Son and Company, 1861); W.T.Sherwin, *Memoirs of the Life of Thomas Paine* (Ondon: R. Calile, 1819), p. 273.

(43) Pierre Cortlandt Van Wyck (1778-1827) was at various times New York City District Attorney, Recorder, Assistant Alderman and prominent member of the Democratic-Republican Party. John Hammond, *The History of Political Parties in the State of New-York, from the Ratification of the Federal Constitution to 1840* (Cooperstown, N.Y.: H & E Phinney, 1846), vol. I, p. 234, p. 238; *American Masonic Record and Albany Saturday Magazine*, Vol. I, p. 78j; Franklin Benjamin Hough, *The New York Civil List*, (Albany, New York: Weed, Parsons and Co., 1860), p. 377, p. 428.

(44) Caesar Augustus Rodney (1772-1824) served in the Delaware General Assembly and State Senate, U.S. House of Representatives, U.S. Senate, and was U.S. Attorney General and Minister to Argentina. Rodney, a nephew of a signer of the Declaration of Independence, was later instrumental in the treason trial of Aaron Burr and helped lay the foundation for the Monroe Doctrine. *DAB*, VIII, pp. 82-83.

(45) A protege of Aaron Burr, John Swartwout (1752-1823; spelling appears, incorrectly, as Swartout or Swarthout in some documents) was a New York City businessman who dealt in paints and dyed woods. As a friend of Burr's, Swartout engaged in a duel with DeWitt Clinton, who had publicly smeared Burr, and was shot in the leg and thigh, leaving him with a permanent limp. Swartout was a New York State Assemblyman, United

States Marshal of New York, and served in the War of 1812 as General of the State Militia on Staten Island and Quartermaster-General of the U.S. Army. After the war he became president of the new City Bank of New York. Joseph Alfred Scoville, *The Old Merchants of New York City* (New York: Carleton, 1863), pp. 249-54, hereafter cited as Scoville; John Flavel Mines, *A Tour Around New York and MY Summer Here, Being The Recreations of Mr. Felix Oldboy* (New York: Harper & Brothers, Publishers, 1893), pp. 288-91; J. Smith Homans, Jr., ed., *The Bankers' Magazine and Statistical Register,* January, 1862, vol. 16, Part 2, p. 670.

(46) Cheetham, who worked as a hatter in England before arriving in the United States in 1798, established himself as editor of the Democratic-Republican paper "The New York American Citizen" with the help of Aaron Burr. He then turned against Burr to support Clinton and Jefferson. A friend and biographer of Thomas Paine, he wrote pamphlets ("View of the Political Conduct of Aaron Burr", "Political Equality and the Corporation of New York", and "Annals of the Corporation of New York") in which he criticized Burr's behavior in the 1800 political campaign, as well as calling for the extension of suffrage for those eligible to vote for assemblymen and reviewing grievances against the Federalist majority on New York City's Common Council. Cheatham served with Myers in the New York State Militia's 3rd Regiment until 1810. *DAB,* II, p. 47; Andrew Burstein, *The Original Knickerbocker: The Life of Washington Irving* (New York: Basic Books, 2007), pp. 33-34; Sidney I. Pomerantz, *New York: An American City 1783-1803: A Study of Urban Life* (New York: Columbia University Press, 1938), pp. 125-26, p. 138, p. 480. Hereafter cited as Pomerantz; Hugh Hastings, ed., *Military Minutes of the Council of Appointment of the State of New York, 1783-1821* (Albany: James B. Lyon, 1901), Vol. II, p. 1158, hereafter cited as "Hastings, *Military Minutes.*"

(47) A physician by profession, Beekman M. Van Beuren (1732-1821) was a lieutenant colonel of the First Brigade of the *New York State Militia.* Rocellus S. Guernsey, *New York City and Vicinity During the War of 1812-15; Being a Military, Civic and Financial Local History of That Period* (New York: C.L. Woodward, 1889), vol. I, p. 101, hereafter cited as Guernsey; Hugh Hastings, ed., *Public Papers of Daniel D. Tompkins, Governor of New York 1807-1817: Military* (New York and Albany: Wynkoop Hallenbeck Crawford, 1898), vol. I, pp. 218-219.

(48) Daniel D. Tompkins (1774-1825), was prominent in New York's Democratic-Republican circles. He became Associate Justice of the New York State Supreme Court by age 30 and Governor of New York State in 1807. While serving as governor he was directly responsible for General Van Rensselaer's army on the Niagara Frontier. DAB, IX, pp. 583-84; Heidler, pp. 516-18.

(49) Colonel Irénée Amelot de LaCroix was a French émigré infantry colonel and former engineer officer who proposed in 1808 that an American military school to be located in Philadelphia and two other cities "at the extremes of this great republic." Madison pressed for de LaCroix's plans but Congress rejected it. The author of *Modern French Tactics or Military Instructions* and *Military and Political Hints, including The Artillerist,* he established a private military school in New York which Myers attended for two years at his own expense. Marcus Cunliffe, *Soldiers & Civilians: The Martial Spirit in America 1775-1865* (Boston & Toronto: Little, Brown and Company, 1968), p. 257, p. 464 n.1; Donald E. Graves, "For want of this precaution so many Men lose their Arms: Official, Semi-Official and Unofficial American Artillery Texts, 1775-1815 Part 8: The Flood Tide of French Inflouence: The Work of Tousard and Duane, 1807-1810." *The War of 1812 Magazine,* Issue 15: May 2011; Donald E. Graves, *Field of Glory: The Battle of Crysler's Farm, 1813.* (Toronto, ON: Robin Brass Studio, 1999), p. 12, hereafter cited as Graves, *FOG.*

(50) "Democratic Clubs" or "Societies" sprang up at this time in many American towns and cities. The French Revolution had inspired French political clubs, and the Democratic-Republicans followed suit, forming networks of similar associations throughout the country. The "New York Club" formed quickly, with most of its members (like Commodore Nicholson, Henry Rutgers and others) coming from the Tammany Society. Tammany was then drawn into city politics and by 1797 had come under the domination of Aaron Burr. The New York Democratic Society corresponded with clubs in other locations, which in turn helped to build up the Democratic-Republican Party. See Pomerantz, pp. 116-17.

(51) John Broome (1738-1810) was a New York City merchant active in the China and India trade who had an active and varied political career. He was a member of the New York Provincial Congress, acted as a city Alderman, City Treasure, President of the New York City Chamber of Commerce and Chairman of the New York City Health Committee during the 1798 yellow fever epidemic. Broome went on to serve in the New York State Assembly and Senate and served as Lieutenant Governor three times under Governors Morgan Lewis and Daniel D. Tompkins. Broome County and the Town of Broome, New York are named for him. Scoville, p.208 ff.

(52) De Witt Clinton (1769-1828) was a mayor of New York City, U.S. Senator, 6th Governor of New York State, and the man largely responsible for the construction of the Erie Canal. *DAB,* II, pp. 221-25; Heidler, pp. 112-13.

(53) Henry Remsen, Sr. (1736-1792) was a New York City merchant and importer who had served in the Continental Congress and was one of the "One Hundred" selected to be in charge of New York City in 1774. Before the Constitution came into effect, John Jay was appointed Secretary of Foreign Affairs by Congress and Remsen was made his Under Secretary. Marion King, *Books and People: Five Decades of New York's Oldest Library* (New York: Macmillan, 1954), p. 216, p. 218; Austin Baxter Keep, *History of the New York Society Library* (New York: De Vinne Press, 1908), p. 33n, p. 203, p. 206.

(54) Peter Philip Schuyler (1776-1825),a native of Albany, NY, served in the *Third* or *Watervliet Regiment* of the Albany County Militia under his father, Philip Peter Schuyler, after which he enlisted in the United States Army. He became a colonel of the *13th U.S. Infantry* on May 12, 1812, then colonel in the Adjutant General's office from April 28, 1813 to June 1, 1814. Schuyler was then assigned to the First Military District (Massachusetts and New Hampshire) and served in his father-in-law's division (Schuyler's first wife was the daughter of Brig. Gen. Thomas Cushing) before being honorably discharged on June 15, 1815. He was then appointed Treasurer of the State of Mississippi and died at Natchez of yellow fever. Francis B. Heitman, *Historical Register and Dictionary of the United States Army From Its Organization, September 29, 1789 To March 2, 1903* (Washington: Government Printing Office, 1903), Vol. I, p. 867, hereafter cited as "Heitman, *HDRUSA*"; Peter Haring Judd, *More Lasting Than Brass: A Thread of Family from Revolutionary to Industrial Connecticut* (Boston: Northeastern University Press, 2004), p. 130, n.77.

(55) First Lieutenant Joseph C. Eldridge, a New York native serving in the *13th U.S. Infantry*, was among 28 American soldiers killed by Native allies under Chief Blackbird in an ambush on July 8, 1813, near Fort George. Eldridge Street in New York City's Lower East Side was named in his memory in 1817. Heitman, *HDRUSA*, Vol. II, p. 21; Graves, *FOG*, pp. 38-41.

(56) Charlotte, Vermont, was named for Charlotte Sophia of Mecklenburg Streliz, wife of George III. The town lies within the fertile Champlain lowlands, a large portion of which was once part of New York. John Duffy, Samuel B. Hand, Ralph H. Orth, *The Vermont Encyclopedia* (Lebanon, New Hampshire: University Press of New England, 2003), p. 83.

(57) Thomas Vail of New York was a Second Lieutenant in the *29th U.S. Infantry*. Heitman, *HDRUSA*, Vol. I, p. 979.

(58) David Curtis, also of New York, was a First Lieutenant in the *15th U.S. Infantry*. Heitman, *HDRUSA*, p. 346.

(59) Skenesborough, now Whitehall, NY, was the Post-Township of Washington County, at the head of Lake Champlain, located about 71 miles from Albany. Horatio Gates Spafford, *A Gazetteer of the State of New York* (Interlaken, New York: Heart of the Lakes Publishing, 1981), Rpt. of 1825 ed., pp. 568-69, hereafter cited as "Spafford."

(60) Willsborough (now Willsboro) was the Post-Township of Essex County, NY on the western shore of Lake Champlain, 138 miles east of Albany and 30 miles south of Plattsburgh. The town was settled by Connecticut farmers, Willsboro had a population of about 900, one blockhouse, one tavern, a Congregational Church, and a grist mill, distillery, ashery, paper mill, iron works, limestone quarry, a cotton and woolen factory, and 5 sawmills. Spafford, p. 573.

(61) Connecticut native John McNeil (1741-1813), a veteran of the French and Indian War, settled in Tinmouth, NY, in 1777. When General Burgoyne invaded New York, McNeil asked for the protection of the British Army, after which his property was confiscated. He was forced to lodge with a known Loyalist, causing suspicion that he, too, was a Loyalist. McNeil moved to Charlotte, VT, in 1784 and became judge of Probate and County Court. In 1791 he established the first toll ferry between Charlotte, VT, and Essex County, NY, from what is still called McNeil's Cove. Most travelers from Western Vermont to Northern New York had to cross the lake at this point, and it became well known as the safest and best ferry on Lake Champlain. E.P. Walton, ed. *Records of the Council of Safety and Governor and Council of the State of Vermont* (Monpelier: Steam Press of J. & J. M. Roland, 1873), Vol. I, pp. 193-94, n.1; Zadock Thompson, *History of Vermont, National, Civil, and Statistical* (Burlington: Chauncy Goodrich, 1842), p. 51; Ebenezer Mack Treman and Murray L. Poole, *The History of the Treman, Termaine, Truman Family in America* (Ithaca, NY: Press of the Ithaca Democrat, 1901), pp. 341-42.

(62) Jones' Hotel was a tavern built in 1809 by Isaac Jones and situated in the downtown area of Willsborough (Willsboro). It was used for early town meetings and school related meetings. There was an active militia at Willsboro during the War of 1812, thought there is no surviving evidence of local recruiting by the regular army. Jones' Hotel was destroyed by fire in 1834-35. Courtesy Ron Bruno, Heritage Society/Willsboro Heritage Center Museum, Willsboro, NY.

(63) Willard Trull, a native of Washington County, NY, was a member of the Washington County militia in 1795. After moving to New York City, he was a commissioned a captain of the *23rd U.S. Infantry* on March 12, 1812, serving mainly as a recruiter. Trull resigned his commission on June 30,

1813 and died at Otsego County, NY on August 28, 1838. Hastings, *Military Minutes*, Vol. I, p. 311, p. 432; Heitman, *HDRUSA*, Vol. I, 972; *Historical Register of the U.S. Army From Its Organization September 29, 1789, to September 29, 1889* (Washington, D.C.: The National Tribune, 1890), p. 451. Hereafter cited as "Heitman, *HRUSA*."

(64) Moses Blatchley (1769-1829) served as an ensign in Lt. Col. Benjamin Strong's Suffolk County New York Regiment in 1793, after which he became a physician and secretary of the Suffolk County Medical Society. Blatchely was commissioned a captain in the *15th U.S. Infantry* on March 24, 1812 but resigned the following September 30th. Ebenezer Prime, *Records of the First Church in Huntington, Long Island, 1723-1779* (Huntington, N.Y., 1899), p. 56; *Scudder Collection of Long Island Genealogical Records*, (Huntington Historical Society), p. 1473; Hastings, *Military Minutes*, Vol. I, p. 282; Heitman, *HDRUSA*, Vol. I, p. 221.

(65) From the time of the Revolution until the start of the War of 1812, Plattsburgh was a town of sawmills, cloth mills, salmon fishing, the rum, timber and pelt trade, and land speculation. In September, 1812, the U.S. Army began constructing 200 log barracks on Lake Saranac, 3 miles upriver from Lake Champlain, which increased the town's population to about 3112 people. General Dearborn's army retreated to Plattsburgh in November, 1812, after encountering Canadian and allied Native troops on their way to attack Montreal. The town was also the scene of the Battle of Plattsburgh, September 11, 1814.

(66) Isaac Clark (1749-1822) began his military career as a First Lieutenant of the *Vermont Rangers* during the Revolution, participating in the Battle of Bennington and the recapture of Fort Ticonderoga. He went on to serve in the Vermont General Assembly and as a Rutland County Judge and on the Council of Censors. During the War of 1812 Clark was a Colonel of both the *11th* and *26th U.S. Infantry Regiments* in the Champlain District. 1200 regulars, including 700 from Boston under Brigadier General John Chandler, had gathered at Burlington, Vermont, at the head of Lake Champlain when Clark became commander of the Cantonment. Clark purchased a 10-acre lot adjacent to the Battery at Burlington and constructed 14 buildings (including barracks, storehouses, gun sheds, magazines a guardhouse and hospital). During the winter of 1812-13, 1/8 of the soldiers at Burlington died of disease, exposure and bad food and water. Clark and Major (later breveted Lieutenant Colonel) Benjamin Forsythe commanded the advance of Wilkinson's army at La Colle Mill, March 30, 1814. Frances B. Heitman, *Historical Register of Officers of the Continental Army During the War of the Revolution, April, 1778 to December, 1783* (Washington, D.C., 1893), p. 156, hereafter cited as "Heitman, *HROCA*"; Heitman, *HDRUSA*, Vol. 1, p. 125;

Benson J. Lossing, *Pictorial Field-Book of the War of 1812* (New York: Harpers, 1868), p. 790. Hereafter cited as "Lossing."

(67) Prior to the war, Helmbold had been the editor of Philadelphia tabloid journal, *The Tickler* in which he named and pilloried both politicians and citizens alike with ridicule and vindictiveness. Lawsuits caused him to enlist in the regular army, first as a sergeant, then ensign in the *13th U.S. Infantry,* and later as a Quartermaster. He was promoted to lieutenant at Lundy's Lane, where he supposedly stirred his men to action by announcing that "those who were born to be hung were in no danger from cannonballs and bullets". After the war he ran the Minerva Tavern in Philadelphia and became editor of *The Independent Balance,* Heitman, *HDRUSA,* Vol. 1, p. 521; John Thomas Scharf, *History of Philadelphia, 1609-1884* (Philadelphia: L.H. Everts & Co., 1884), Vol. III, pp. 1983-86; *The Constitution and Register of the membership of the General Society of the War of 1812* (Philadelphia: 1908), p. 193, hereafter cited as *"CRMGS War of 1812."*

(68) This individual cannot be identified.

(69) Plattsburgh, New York.

(70) Greenbush, New York, on the east bank of the Hudson River directly across from Albany, was chosen as the Northern Army's Military Cantonment by General Henry Dearborn in May, 1812. The first troops to arrive there were quartered in tents but were eventually housed in 24 buildings. During July, August and September of 1812, the 1500-3000 men quartered there suffered from dysentery, measles, pleurisy, fatigue, poor sanitation and bad food and water. By September-November 1, 1812 most of Dearborn's army, which had increased to 6000-8000 regulars and militia, had marched to the Niagara Frontier. *Description of Barracks at the Cantonment at Greenbush,* Record Group No. 94, AGO Miscellaneous File, the National Archives, Washington, D.C.; Nathaniel Bartlett Sylvester, *History of Rensselaer County* (Philadelphia: Everts & Peck, 1880), pp. 333-337, p. 352, pp. 358-59; George Rogers Howell and Jonathan Tenney, *History of the Count y of Albany, N.Y., From 1609 To 1886* (New York: W.W. Munsell & Co., Publishers, 1886), pp. 493-94; *Greenbush Cantonment, Town of East Greenbush – Clinton Heights, N.Y.,* unpublished manuscript, n.d., courtesy of the East Greenbush Historical Society; Kay Mowers, "Cantonment Farm," *Historic Albany Area No. 15,* December 17, 1941. For the troops' living conditions while at Greenbush, see James Mann, *Medical Sketches of the Campaigns of 1812, 13, 14* (Dedham, MA: 1816), et passim.

(71) Lt. Col. Stephen Thorn began as an ensign in the 3rd Company, Hoosack and Schaghtecooke, under Col. John Knickerbocker. At the start of the War of 1812 he was Commandant of the *14th Regiment of Artillery* of the New York State Militia. Guernsey, Vol. 1, p. 101; *CRMGS War of 1812,*

p. 193; Berthold Fernow, *Documents Relating To The Colonial History of the State of New York* (Albany: Wood Parsons & Co., Printers, 1887), Vol. 1, n.p.

(72) Massachusetts native Benjamin H.Mooers (1758-1838) was a veteran of the American Revolution, settled in Plattsburgh in 1783 and served 8 years in the New York State Assembly. Mooers was made First Lieutenant of the *23rd U.S. Infantry Regiment* on May 15, 1812, resigned June 13, 1813, and became Major General of the New York State Militia. He commanded a body of soldiers at the Battle of Plattsburgh, September 11, 1814. "Biography of Major-General Benjamin Mooers of Plattsburg, Clinton County, N.Y., Written in 1833, By Request of His Son, Benjamin H.Mooers." *The Historical Magazine and Notes and Queries Concerning the Antiquities of History and Biography of America*, Vol. I, third Series (Morrisania, N.Y.: Henry B. Dawson, 1872-73), pp. 92-94; Heitman, *HDRUSA*, Vol. I, 721; Lossing, p. 875, n.4. While camped with Mooers at Cumberland Head, Myers saved the life of a drunken soldier (unnamed) by pulling him out of a campfire on the beach. The soldier visited Myers at his land office on Wall street years later to reminisce about the incident. See Myers's *Reminiscenses*, p. 51.

(73) Cumberland Head, part of Plattsburgh, NY, is a peninsula projecting into Lake Champlain.

(74) See Benjamin Mooers to Mordecai Myers, September 4, 1812. Theodorus Bailey Myers Collection, #1852, New York Public Library, hereafter cited as TBMC; Benjamin Mooers to Mordecai Myers, September 5, 1812. Mordecai Myers Collection, The Clements Library, University of Michigan. Hereafter cited as MMC.

(75) Edward Baynes (? – 1829) became a Lieutenant-Colonel of the *Glengarry Light Infantry Fencible Regiment* in 1812, based in Upper and Lower Canada, and served briefly as military secretary to Isaac Brock. As Adjutant-General of British North America during the war, he conducted the negotiations of the Prevost-Dearborn Armistice of 1812 when he went to Albany under a flag of truce. He was one of General Sir George Prevost's subordinate officers and "commanded the 750-man force ordered by Prevost to take Sacket's Harbor in May 1813." Heidler, pp. 42-43.

(76) Henry Dearborn (1751-1829) was a physician, veteran of the Revolution, Democratic-Republican Congressman (1793-97), Secretary of War (1801-1809) and Collector of the Port of Boston (1809-1812). Dearborn was appointed senior Major General in the U.S. Army in January, 1812, to command the northeast sector from the Niagara River to the New England coast. He was recalled July 6, 1813, for ineffective leadership, reassigned to an administrative command in New York City and honorably discharged on June 15, 1815. Madison appointed him Minister Plenipotentiary to

Portugal in 1822, and he returned to Boston 2 years later. William Gardner Bell, *Commanding Generals and Chiefs of Staff: Portraits and Biographical Sketches* (Washington, D.C.: Government Printing Office, 1992), pp. 72-73; *DAB*, vol. III, 174-76; Heidler, pp. 146-47; Heitman, *HROCA*, p. 190; *HDRUSA*, Vol. I, p. 363.

(77) The village of Burlington, Vermont, played a vital role in the defense of the lakes during the War of 1812 as a staging ground for unsuccessful attacks against Canada. The facility was abandoned in 1817. See Note 66, and James P. Millard, *Burlington, Vermont, during the War of 1812* (www.historiclakes.org/explore/burlington.htm).

(78) West Point graduate Samuel B. Rathbone of New York was a Second Lieutenant in an artillery company and served in various Atlantic posts. Rathbone died on December 8, 1812, of wounds received at the Battle of Queenston Heights, October 13, 1812. Heitman, *HRUSA*, p. 540; George W. Cullum's *Biographical register of the officers and graduates of the U.S. Military Academy, at West Point, N.Y., from its establishment, March 16, 1802, to the army re-organization of 1866-67. By Bvt. Major-General George W. Cullum.* (New York: D. Van Nostrand, 1868), Vol. I, p. 87. Hereafter cited as "Cullum."

(79) Winfield Scott (1786-1866), known as "Old Fuss and Feathers" and the "Grand Old Man of the Army," was the longest serving (47 years) General in American history. Scott fought in the War of 1812, the Mexican-American War, the Black Hawk War, the Second Seminole War, and was commanding General of the U.S. Army at the start of the the American Civil War. As Brigadier General he supervised the removal of the Cherokee, westward across the Mississippi in 1838 (the "Trail of Tears") and ran unsuccessfully for President as a Whig in 1852. He was court-martialed and suspended for one year for criticizing General James Wilkinson (Scott was then a Captain), led a landing party at Queenston Heights (where he was captured and paroled), led the capture of Ft. George (where he was wounded), and was wounded again at the Battles of Chippawa and Lundy's Lane. Heidler, pp. 464-65; *DAB*, VIII, 505-511.

(80) Located 3 miles east of Buffalo, the Flint Hill Encampment was the area of the main road between Conjackety (Scajaquada) Creek and present Jewett Parkway. Three hundred soldiers died at Flint Hill from exposure, disease and inadequate food during the winter of 1812-13.

(81) Alexander Smyth (1765-1830) had served in the Virginia House of Delegates and Senate before receiving a political appointment to the U.S. Army first as a colonel, inspector-general, then brigadier general in 1812. Though he had no practical military experience, he had authored a pamphlet of field maneuvers. As an officer in the regular army, he refused to

support General Stephen Van Rensselaer, commanding the New York State militia at Queenston Heights. Smyth left the Niagara Frontier in disgrace after two failed attempts to invade Canada, but served in the Virginia House of Delegates and the U.S. Congress after the war. Heidler, 479-81; Robert S. Quimby, *The U.S. Army in the War of 1812: An Operational and Command Study* (East Lansing, Michigan: Michigan State University Press, 1997), Vol. I, pp. 77-78, hereafter cited as "Quimby"; John R. Elting, *Amateurs, To Arms! A Military History of the War of 1812* (Chapel Hill, North Carolina: Algonquin Books, 1991), p. 51, hereafter cited as "Elting."

(82) Buffalo, NY, saw more of the War of 1812 than any other town in the U.S., though its population at the time was no more than 400. Buffalo guarded the eastern end of Lake Erie where it flowed into the Niagara River. Fort Erie, a British post, was directly across the Niagara's southern mouth. General Alexander Smyth had encamped his troops nearby at Flint Hill; after the Americans burned Newark, the British burned Buffalo; Winfield Scott's brigade came to Buffalo in April, 1814; it became the base for General Jacob Brown's invasion of Canada; and, reinforcements and supplies from Buffalo sustained U.S. troops holding the captured Fort Erie during its 6-week siege. After the war, Buffalo became the western terminus of the Erie Canal and the starting point for westward expansion.

(83) In the fall of 1812, Commodore Isaac Chauncey took command of Lakes Erie and Huron, and sent Lt. Jesse Elliott to Lake Erie to build ships for service. Elliott found out that 2 of the 5 ships on the lakes, the *Detroit* (formerly the American ship *Adams*) and the *Caledonia*, were under the guns at Fort Erie and decided to take them. Winfield Scott, Nathaniel Towson and 100-124 Americans in 2 schooners left Buffalo/Black Rock and rowed toward Fort Erie. Early on October 9th, Elliott and Towson's men took the 2 brigs; after 10 minutes of fighting, the ships were allowed to drift in the hopes that they would reach the American shore. The *Detroit* drifted onto Squaw Island and was finally burned by the Americans to prevent its recapture, while the *Caledonia* was saved and became the "nucleus of the American flotilla on Lake Erie". John K. Mahon, *The War of 1812* (Gainesville, Florida: The University Presses of Florida, 1972), pp. 90-91, hereafter cited as "Mahon"; Donald R. Hickey, *The War of 1812: A Forgotten Conflict* (Urbana and Chicago: University of Illinois Press, 1989), p. 131. Hereafter cited as "Hickey"; Quimby, vol. I, pp. 66-67; Heidler, p. 54.

(84) Jesse Duncan Elliott (1782-1845) was a Maryland-born orphan who served as midshipman on the "Essex," lieutenant on the "Chesapeake", and acting lieutenant on the "Enterprise". When war was declared he went to New York City and was then sent to Presque Isle on Lake Erie in Pennsylvania to build a fleet of ships. Elliott was one of the principal officers in the "Caledonia Incident" of October 9, 1812. He was promoted to master

commander in 1813 and was with Chauncey during the attack on York and the Battle of Lake Erie. Elliott was given a 4-year suspension without pay after a trial for misconduct, but this was settled 2 years later and he became commandant of the Philadelphia Navy Yard in 1844. Heidler, p. 165.

(85) Lt. Col. Thomas Miller, a native of Plattburgh, NY, was the commandant of the *8th Regiment* of the New York State detached militia, which came under Captain Ezra Turner's *36th Infantry Battalion* on July 20, 1813. Miller was in the U.S. service during Murray's Raid on Plattsburgh (July 29-August 5, 1814) and the Battle (or Siege) of Plattsburgh, on September 14, 1814. His home in Plattsburgh is now an historic site.

(86) Squaw Island, was called "De-dyo-we-no-guh=doh" by the Senecas, referring to its divisions by the marshy creek called "Smuggler's Run." Squaw Island was the site of troop landings and departures, as well as the spot where the *Detroit* was beached, burned and sunk by the Americans on October 9, 1812.

(87) Williams, whom Myers's granddaughter referred to as her grandfather's "faithful servant", was born in Pennsylvania sometime in 1790-96. He was a private in the *19th Infantry Regiment* from Washington (or Allegany) County, PA, which also included soldiers from Ohio under Captain William A. Trimble. During the summer of 1814, the *19th* was ordered to the Niagara Frontier under Major General Jacob Brown and participated in the capture of Fort Erie and the Battle of Lundy's Lane. The regiment was later placed in the Second Division under Colonel James Miller. Williams died on September 3, 1814. *U.S. Army Register of Enlistments 1798-1814*, Entry #5748, National Archives, Washington, D.C.; Heitman, *HDRUSA*, Vol. I, p. 970.

(88) Amos Hall (1761-1827), a native of Connecticut, was a 15-year-old fifer in his father's regiment during the Revolution and later helped settle the town of Bloomfield, NY. He took the first census of Western New York and served as a U.S. Deputy Marshall, New York State Assemblyman and Senator, and Commander-in-Chief of the Niagara Frontier until replaced by Major General Stephen Van Rensselaer on August 11, 1812. Hall served as an officer under militia General George McClure until McClure's removal in December, 1813 when, on Christmas Day, McClure handed his command to Hall. Hall then established headquarters between Black Rock and Buffalo in response to differing reports of an impending British attack. When the disorganized militia units failed to guard the bridge near Shogeoquady Creek, they tried to halt their retreating comrades from running through the streets of Buffalo. Hall was subsequently replaced by General Peter S. Porter as the commander of militia on the Northern

Frontier. He died in Ontario County, NY in 1827. Heidler, 222-23; "Hall's Report to Governor Tompkins" in Lossing, 635, n.4.

(89) Myers may be referring to the *46th Regiment* which, along with the *4th, 9th, 13th, 21st* and *40th Regiments*, were consolidated into the new *5th Regiment*, organized on May 15, 1815, under the command of Colonel James Miller.

(90) Colonel Asa Stanton (1760-1817) was a farmer and tavern keeper from Norwich, CT. During the American Revolution he was a private, then captain in the Connecticut militia, participating in the Battle of White Plains and subsequently imprisoned on the "Jersey" in New York Harbor. Stanton relocated to Waymart, Wayne County, PA in 1790 where he built a sawmill on Stanton's Pond and was elected a deputy sheriff, captain, then colonel of the local militia in 1812. After the war he moved to Bethany, PA where he drowned in a boating accident in 1817. Tom Carney, "Col. Asa Stanton", www.finda grave.com.

(91) Lt. Col. Thompson Mead (1774-1851) of Dutchess County, NY, commanded the 400-man *17th Regiment of New York Militia* from early September 1812. The regiment was taken prisoner at Queenston and taken first to Niagara, then Newark. Mead was later promoted to general of the New York State militia. Hiram C. Clark, *A History of Chenango County* (Norwich, N.Y.: Thompson & Pratt, 1850), pp. 107-113; Spencer Percival Mead, *History and Genealogy of the Mead Family of Fairfield County, Connecticut, Eastern New York, Western Vermont and Western Pennsylvania, from A.D. 1180 to 1900* (New York: The Knickerbocker Press, 1901), p. 91; Lossing, p. 366, n.2; John Ward Dean, George Folsom, John Gilmary Shea, *The Historical Magazine and Notes and Queries Concerning The Antiquities, History and Biography of America*, Vol. II, third Series (Morrisania, N.Y.: Henry B. Dawson, 1873), p. 99.

(92) Brigadier General Peter Buell Porter (1773-1844), a graduate of Yale, was a lawyer, soldier and politician who practiced law at Canandaigua, NY. After serving as clerk of Ontario County and in the New York State Assembly, Porter moved to Black Rock and was elected as a Democratic-Republican Congressman. He was made Quartermaster-General of the New York State Militia (May-October, 1812), participated in General Smyth's abortive attempts to invade Canada and fought a "bloodless duel" with the general. He raised and commanded a brigade of New York militia that included a Six Nations contingent, which he led with distinction. Porter received a gold medal from Congress on November 3, 1814, "for gallantry and good conduct" during the Battles of Chippawa [a.k.a. Chippewa], Niagara [a.k.a., Lundy's lane, and Niagara Falls] and Erie [Fort Erie]. Elting, 51.

(93) Jesse Duncan Elliott. See also, Note 83.

(94) "Sprowell" was John Sproul of New York served with Myers in the state militia, first as an Ensign. He became a Captain in the *13th U.S. Infantry* on March 12, 1812, then Major on July 25, 1814. Sproul was retained as Captain of the *2nd U.S. Infantry* on May 17, 1815, with brevet of Major from July 25, 1814. He resigned from the army on May 1, 1819. Heitman, *HDRUSA*, vol. I, 913; Hastings, Military Minutes, vol. II, 1158.

(95) Hugh R. Martin (1771-1848) was a native of Schenectady, NY. He was commissioned a Captain in the *13th U.S. Infantry* on March 12, 1812, Major of the *22nd Infantry*, September 12, 1814, and honorably discharged June 15, 1815. After the war, Martin returned to Schenectady where he operated a successful brewery. Heitman, *HRUSA*, p. 692; Hugh Hastings, ed., *Public Papers of Daniel D. Tompkins, Governor of New York 1807-1817: Military* (New York and Albany: Wynkoop Hallenbeck Crawford, 1898), Vol. II, pp. 411-412; Schenectady *Reflector*, May 19, 1848, p. 2.

(96) Thomas Humphrey Cushing (1755-1822) was a veteran of the American Revolution. He remained in the army and was promoted to Colonel, Adjutant-General, Inspector-General, and Brigadier General, supervising the construction of forts in the Mississippi Territory and commanding several posts west of the Mississippi. Cushing was court-martialed at Baton Rouge in 1811 on charges of disobedience of orders, abuse of trust and misapplication of public property, and "Conduct Unbecoming an Officer and a Gentleman". He was found not guilty, reprimanded and released. In 1816 he was appointed Collector of Customs for the port of New London, CT. A year later he fought a duel with Virginia Congressman William J. Lewis and was saved when the bullet struck his watch. Cushing died at New London on October 19, 1822. Myers was incorrect. Peter Philip Schuyler's first wife was Cushing's adopted daughter. Cushing was Schuyler's father-in-law. *Historical Collections – Collections and researches Made By the Michigan Pioneer and Historical Society* (Lansing, Michigan: Wynkoop Hallenbeck Crawford Co. State Printers, 1905), Vol. XXXIV, p. 374, n.8; *Annual Reports of the War Department For the Fiscal Year June 30, 1900. Reports of Chiefs of Bureaus* (Washington: Government Printing Office, 1900), p. 223; James S. Cushing, *The Genealogy of the Cushing Family* (Montreal: The Perrault Printing Company, 1905), pp. 142-43; *Appleton's Cyclopedia of Biography* (New York: D. Appleton and Company, 1888), p. 221.

(97) Black Rock, once an independent municipality, is now a neighborhood of the northwest section of Buffalo. Its name came from the large outcropping of black limestone along the Niagara River which was blasted away in the 1820s to make way for the Erie Canal.

(98) Willoughby Morgan (mid-1780's-1832) was the illegitimate son of Daniel Morgan (1736-18020), the Revolutionary War leader of "Morgan's Rifles" or "Morgan's Riflemen". Morgan was commissioned a Captain in the *12th U.S. Infantry* on July 6, 1812, and later rose to Lieutenant Colonel, commanding several outposts in Missouri, Nebraska, and Wisconsin, and negotiating treaties with several of the western tribes. Heitman, *HDRUSA,* vol. I, p. 81; *DAB,* VIII, pp. 166-67; Dan Higgenbotham, *Daniel Morgan: Revolutionary Rifleman* (Chapel Hill, North Carolina: University of North Carolina Press, 1961), p.183.

(99) Massachusetts native and veteran of the Revolution, Moses Porter (1756-1822) commanded the artillery bombardment of Fort George and was the senior artillery officer in Wilkinson's army. His nickname "Old Blowhard" came from his reputation as "the most profane officer" in the service. Heidler, pp. 422-23; Donald E. Graves, "The Hard School of War: A Collective Bibliography of the General Officers of the United States Army in the War of 1812." *The Napoleon Series – Military Subjects: The War of 1812 Magazine,* Issue 3. Part II: "the Class of 1813". June, 2006. Hereafter cited as Graves.

(100) Williamsville is a village in Erie County, NY, now located mostly in the town of Amherst. It was originally called "Williams Mills" after Jonas Williams, who built the first mill there in 1811. During the War of 1812, American troops were stationed there between what is now Garrison Road and Ellicott Creek, and both American soldiers and British prisoners were treated there in a field hospital and log barracks that lined Garrison Road. A small cemetery was used to bury the 205 soldiers who died there. Winfield Scott used the Evans House as his headquarters in Spring 1813, when his army of 5-6000 men were stationed at Williamsville. Sue M. Young, *A History of Amherst, N.Y.,* n.d., Chapters 2-3.

(101) Myers means mid-to-late November of 1812. By November 18th, Smyth had 6000 regulars and militia with which to attempt a two-pronged attack to prepare the way for a larger American invasion by trying to take the Red House and Fort Erie. Myers was among the men of the *13th U.S. Infantry* who were to cross the Niagara with 150 troops and 70 U.S. sailors under Lt. Samuel Angus and spike the batteries at the Red House, 2 and 1/2 miles from Fort Erie. On November 20th, Smyth ended an armistice arranged by General Van Rensselaer, so General Sheaffe had Fort Niagara bombarded from 7:30 A.M. until dark. The Americans at Fort George responded, but Newark was set ablaze and the mess hall at Fort Niagara was damaged, causing about 12 American casualties. In the early hours of November 28th, Americans under Colonel William King overwhelmed the *49th Regiment* and spiked the guns, but in the confusion that accompanied

the return to their landing point, Angus re-crossed the river stranding King's men; he and 38 others were captured. Lt. Col. Charles Boerstler left Lt. John Waring to dismantle the bridge over Frenchman's Creek, but when it was learned that a large force of British from Fort Erie was coming, Boerstler left for Buffalo, leaving Waring and his party at the bridge. The "Battle of Frenchman's Creek" or "the Affair opposite Black Rock" was an American failure. Smyth tried again on November 31st, but by the time less than half of his force was loaded onto the boats it was daylight and he postponed the action. His troops became increasingly violent and unruly; Smyth requested and received leave to visit his family in Virginia, and 3 months later he was dropped from the army's rolls. Louis L. Babcock, *The War of 1812 on the Niagara Frontier* (Buffalo, New York: Buffalo Historical Society, 1927), pp. 59-76; James Hannay, *History of the War of 1812 Between Great Britain and The United States of America* (Toronto: Morang & Co. Limited, 1905), p. 90.

(102) These *Pennsylvania Volunteers* were commanded by General Adamson Tannehill (1750-1820), a veteran of Daniel Morgan's Rifle Company in the Revolution, and were part of the *Westmoreland County (PA) Militia* formed just before the war. In a General Order issued from Williamsville, NY, on December 8th, 1812, Tannehill reported in his return that 5 captains, 4 lieutenants, 11 ensigns, 83 sergeants, 89 corporals, 25 musicians and 930 privates in his brigade had "revolted and deserted", leaving only 267 privates on duty. Tannehill requested and was granted leave for himself and his staff, handing the command over to Major Harriett. Ernest A. Cruickshank, ed., *The Documentary History of the Campaign Upon the Niagara Frontier In The Year 1812, Part II* (Welland, Ontario: The Lundy's Lane Historical Society, n.d.), p. 291.

(103) General William H. Winder (1775-1824) had a career in the law and politics before the war, serving in the Maryland legislature. He was commissioned a Lieutenant Colonel, then Colonel in the infantry, and commanded the *14th U.S. Infantry Regiment* "which he led to the Niagara frontier in the Winter of 1812-13." After his promotion to Brigadier General in March 1813, he commanded a brigade in the assault on Fort George (May 27, 1813) "but was captured in early June, along with Brigadier-General John Chandler at the Battle of Stoney Creek." After his release in Spring 1814 he became Adjutant Inspector General, then placed in command of the Tenth Military District, where he tried but failed to defend Washington and Balitmore. Despite a ruined constitution, court-martial and discharge in 1815, Winder was twice elected Senator from his home state and served as the Grand Master of the Masonic Order of the State of Maryland. Heidler, p. 88, pp. 558-60; *DAB*, X, pp. 382-83; Heitman, *HDRUSA*, vol. I, 1049; Lossing, p. 918, n.2.

(104) Lt.-Col. Christopher Myers commanded 750 troops of the *100th Foot* during the withdrawal from Crooks Mills/Lyons Creek on October 19, 1814, the last engagement on the Niagara Frontier. Myers had attacked outposts across the Creek that day but saw he was outnumbered by the Americans under General Daniel Bissell's 900-man brigade. Heidler, pp. 127-28.

(105) Gen. Charles Boerstler to Myers from Buffalo, February 23, 1813. *MMC*, Clements Library; Myers to Naphtali Philips from Williamsville, *PAJHS*, XXVII, pp. 396-97; See Myers's order for provisions written from Williamsville, April 8, 1813. *MMC*, Clements Library.

(106) New Jersey native Zebulon Montgomery Pike (1779-1813) came from a military family and was a Colonel by 1812. He marched 600 men of the *15th U.S. Infantry Regiment* from Greenbush to Plattsburgh in October 1812 to participate in Major General Henry Dearborn's aborted offensive against Montreal. After promotion to Brigadier General on March 12, 1813, Pike commanded a brigade in the attack on York (April 27, 1813) but was killed "by debris from the explosion of a British magazine." He explored the Mississippi and New Mexico Territories. Pike's Peak is named in his memory. Graves, pp. 6-7; Heidler, pp. 415-416; *DAB*, VII, pp. 599-600.

(107) William Hull (1753-1825), a veteran of the Revolution, was appointed Governor of Michigan by Thomas Jefferson in 1805. Hull surrendered Detroit to the British on August 16, 1812.

(108) Welsh-born Henry Procter (1763-1822) was made a Major-General for his victory at Frenchtown under General Brock, which prevented Hull's invasion of Canada. Procter was defeated by General William Henry Harrison at the Battle of the Thames (October 5, 1813), casusing the death of the Shawnee chief and British ally Tecumsah. A. M. J. Hyatt, "Henry Procter." In *Dictionary of Canadian Biography Online*. Vol. 6, 1821-1835. Toronto: University of Toronto Press, 1979. Hereafter cited as *DCBO*.

(109) New Hampshire attorney James Miller (1776-1851) entered the army as a Major in 1808. As a Lieutenant Colonel he commanded the Americans at the Battle of Brownstown, under William Henry Harrison at Tippecanoe, and was recognized for distinguished service near Detroit. Miller commanded the *6th Infantry Regiment* on the Niagara Frontier in May 1813; in March 1814 he commanded the *21st Infantry Regiment* as full Colonel under General Jacob Brown at Chippawa, Niagara Falls and Fort Erie. He was promoted to Brigadier General for capturing the battery at Lundy's Lane, for which Congress gave him a gold medal. Miller left the army in 1819 and became Governor of the Arkansas Territory and Collector of the Port of Salem. He died at Temple, NH, in 1851 after surviving 2 strokes. Lossing, pp. 819-820, n.1; Heidler, p. 67, p. 171, p. 248, p. 336, pp. 352-53, p. 522.

(110) Bostonian Josiah Snelling (1782-1828) was the son of a prominent banker who began his military career as a First Lieutenant (permanent rank commission) in the *4th Infantry Regiment* on May 3, 1808; regimental paymaster April 5 to June 12, 1809; Captain (permanent rank promotion), June 12, 1809, to February 21, 1814; Brevet Major on August 9, 1812, for distinguished service at the Battle of Brownstown; Lieutenant Colonel (permanent rank promotion) of the *4th Regiment of Riflemen,* February 21, 1814, to May 17, 1815; Lieutenant Colonel of the *6th Infantry Regiment,* May 17, 1815; then Colonel (permanent rank promotion) of the *5th Infantry Regiment* from June 1, 1819, until his death on August 20, 1828. Snelling also fought at Fort Detroit and Tippecanoe. He was the first commander of Fort Snelling, at the confluence of the Mississippi and Minnesota Rivers in Minnesota. Heitman, *HRUSA,* p. 906.

(111) Rhode Island native Benjamin Watson was a Second Lieutenant, First Lieutenant, Adjutant, then Captain of the *25th U.S. Infantry* between March 12, 1812, and August 15, 1813. He was transferred to the *6th U.S. Infantry* on May 17, 1813, and was made a Brevet Major in July 1814 for gallant conduct in the Battle of Niagara Falls (Lundy's Lane). Watson was transferred to the *5th U.S. Infantry* on January 1, 1821. He died in 1827. Heitman, *HDRUSA,* vol. I, 1009.

(112) During the War of 1812, Kingston, Ontario, was the base for the Lake Ontario division of the Great Lakes British Naval fleet which competed with the American fleet based at Sackett's Harbor, NY, for control of Lake Ontario.

(113) The unsuccessful Battles of Chateauguay (October 26, 1813) and Crysler's Field (November 11, 1813; a.k.a. Crysler's Farm) were America's two failed attempts to capture Montreal, which would have led to the conquest of all Upper Canada.

(114) The St. Regis Mohawk Reservation, or *Akwesasne,* now contains the villages of Hogansburg and St. Regis, the latter named for the Catholic saint. Under the terms of the Jay Treaty (1794), the Mohawk People could pass freely across the International Boundary which divides the territory north and south. The Catholic American Iroquois remained neutral during the War of 1812, but a battle between Americans and British in October 1812 caused some native homes to be ransacked. On November 23, 1812, British regulars, Canadian militia and St. Regis Mohawks crossed the border and defeated the Americans in a skirmish at French Mills now Fort Covington, NY. "St. Regis", in Frederick J. Seaver, *Historical Sketches of Franklin County and Its Several Towns With Many Short Biographies* (Albany: J.B. Lyon Company, Printers, 1918), Chapter XXIII.

(115) Thomas McDonough (1783-1825) led his "undergunned and outnumbered flotilla to a crushing victory against a British squadron" under Captain George Downie at Plattsburgh Bay on September 11, 1814. McDonough's victory prevented northern New York from invasion and made the British more flexible to accepting American demands during the peace talks at Ghent. Heidler, pp. 311-14.

(116) A native of Delaware, Jacob Jones (1768-1850) commanded the *U.S.S. Mohawk* on Lake Ontario. He had studied medicine and law prior to joining the new U.S. Navy in 1799. During the Barbary War (1801-1805) he was captured by Tripolitan pirates and held as a POW for 19 months before being released. He went on to command the sloop *Wasp*, and in 1812 defeated *H.M.S. Frolic* in a vicious battle that left both ships disabled. Jones was captured, exchanged, promoted to Captain, and received a Congressional Gold Medal along with command of the recently captured British frigate *Macedonia.* Heidler, pp. 269-70.

(117) Oliver Hazard Perry (1785-1819) led the American Navy to a victory during the Battle of Lake Erie (September 10, 1813), for which he received a Congressional Gold Medal and Thanks of Congress. Perry's victory allowed William Henry Harrison to invade Canada, capture Amherstburg and Detroit, and defeat the British forces at Moraviatown. Heidler, pp. 290-93; pp. 412-14.

(118) Major Isaac Shelby (1750-1826) of Maryland became famous during the American Revolution for his actions at the Battle of King's Mountain (October 7, 1781). He was the first Governor of Kentucky (1792) and on October 5, 1813, directed reinforcements to back Richard M. Johnson's attack against the British and Indian positions. This resulted in the death of Tecumseh and the dispersal of his warriors. General Procter retreated, and over 600 of his men were made prisoners. Congress awarded Shelby a gold medal in 1818. Heidler, pp. 471-72.

(119) Tecumsah was killed at the Battle of the Thames on October 5, 1813, not by Johnson, as was originally believed at the time, but by William Whitley, a veteran of the American Revolution. Heidler, pp. 504-505.

(120) Isaac Chauncey (1772-1840), a Connecticut native, was appointed to command at Lakes Ontario and Erie in September 1812 following Hull's surrender at Detroit. He arrived at Sackett's Harbor with 700 sailors and shipwrights, supported Dearborn's attack on York (April 27, 1813), and landed Dearborn's army to capture Fort George on May 27, 1813. Heidler, pp. 90-91.

(121) Sackett's Harbor is a village in Jefferson County, NY, at the eastern end of Lake Ontario on the south shore of Black River Bay, about 1 mile

from its mouth and 10 miles west and south from Watertown. It was first settled in 1801 by Augustus Sacket, incorporated as a village in 1814, and was the American base for operations on the Great Lakes during the War of 1812. In July of 1812 it was attacked by a Canadian provincial squadron, and again in May 1813 by a British squadron under Commander Yeo and troops under General Sir George Prevost. General Zebulon Pike, who was killed at York (Toronto) on April 27, 1813, is buried in the military cemetery at Sackett's Harbor.

(122) Myers refers here to the "Battle of York (Toronto)" of April 27, 1813. An American force of 1700 troops under General Zebulon Pike, supported by Chauncey's fleet, landed on the Lake Ontario shore to the west. After encountering General Roger Sheaffe's 700 British and Native troops, they captured York and its dockyard. During the next two days they carried out several acts of arson and looting, blowing up ships and naval stores, burning the Legislative Assembly and military buildings, destroying the Printing Office and looting empty houses. Pike and 38 American soldiers were killed when a powder magazine exploded. As York was the capital of Upper Canada at the time, the British retaliated in 1814 by burning Washington. The American victory at York helped the U.S. maintain balance on Lake Ontario and hampered British operations on Lake Erie. Quimby, vol. 1, pp. 225-30; Hickey, pp. 129-30.

(123) The Battle of Queenston Heights (October 13, 1812) in the present Province of Ontario was the first major battle of the War of 1812. Major General Stephen Van Rensselaer led 6000 U.S. regulars and New York State militia forces against 1300 British regulars, Canadian militia and Mohawk allies led by Major-General Sir Isaac Brock, and Major General Roger Sheaffe who took command when Brock was killed. The Americans were attempting to get a foothold on the Canadian side of the Niagara River before the winter. British artillery, the reluctance of undertrained and inexperienced New York militia to cross the river, and the arrival of British reinforcements forced the Americans who had already landed on the Canadian side to surrender. Elting, pp. 41-47; Hickey, pp. 86-88; Quimby, vol. 1, pp. 65-66; J. Mackay Hitsman, *The Incredible War of 1812: A Military History*. Updated by Donald E. Graves (Toronto: Robin Brass Studio), originally published by University of Toronto Press, 1965, pp. 83-92.

(124) Lt. Col. Thompson Mead. See Note #91.

(125) Farrand Stranahan (1778-1826), a native of Connecticut, was commanding the Sixth Brigade (one of 8 Brigades), *16th Regiment* of the New York State Militia when he was taken prisoner by British forces at the Battle of Queenston. Stranahan later became a New York State Senator and member of the Council of Appointment, and was one of the U.S.

Senators who voted against giving the Electoral College directly to the American people. He died at Cooperstown, NY, on October 22, 1826. Henry Reed Stiles, *Genealogies of the Stranahan, Josselyn, Fitch and Dow Families in North America* (Brooklyn, N.Y.: H.M. Gardner, Jr. Printers, 1868), pp. 19-20; Lossing, p. 366, n. 2.

(126) Massachusetts native Decius Wadsworth, who received an M.A. from Yale and was a member of Phi Beta Kappa, was commander of a Regiment of Artillerists and Engineers from January 9, 1800, to April 1, 1802. Wadsworth was appointed Commissary-General of Ordnance to the Ordnance Department established by Congress on May 14, 1814, and charged with inspecting and approving all ordnance pieces, cannonballs and shells, and directing the construction of carriages and other apparatus for field and garrison service. He was designated Chief of Ordnance on Febryuary 8, 1815. Wadsworth died in 1821. *Catalogue of Members Phi Beta Kappa.* (New Haven, Connecticut: Connecticut Alpha [Yale University], 1905), p. 26.

(127) Myers is mistaken. He meant Solomon Van Rensselaer (1774-1852), born in East Greenbush, NY. After enlisting in the army he became Captain, then Major, of a volunteer company by January 8, 1799, and left the army in June, 1800. Van Rensselaer then served as Adjutant-General of the New York State Militia. As Lieutenant Colonel of New York Volunteers he was wounded at the Battle of Queenston. Stephen Van Rensselaer had secured Solomon (his second cousin) an appointment as his aide-de-camp because he was an experienced soldier who had been wounded at the Battle of Fallen Timbers in 1794 and could offer valuable advice; Stephen Van Rensselaer had never commanded troops in battle. After the war, Van Rensselaer was a U.S. Representative from New York (1819-1822) and served as Postmaster of Albany, NY (1822-23, 1841-43). *DAB*, X, pp. 210-211.

(128) Stephen Van Rensselaer (1764-1839), one of America's wealthiest men of the time, enjoyed a varied career as a Major and Major General of the New York State Militia, New York State Assemblyman and State Senator, and Lieutenant Governor. In addition, he was the founder of Rensselaer Polytechnical Institute (RPI) in Troy, NY. Van Rensselaer, a Federalist and leading opponent of the Democratic-Republican Daniel D. Tompkins, was offered command of the U.S. Army of the Center by Tompkins, despite being untrained and inexperienced. Tompkins knew Van Rensselaer would not refuse the appointment for fear of losing public esteem; if he accepted the command, he would be out of the running for Governor; if he turned out to be a good General the military powers would not remove him, and, if a poor one, he would be discredited. Van Rensselaer was defeated at Queenston when the militia refused to cross into Canada

and Alexander Smyth refused to obey him because he was militia rather than regular army. Myers mistakenly refers to Solomon Van Rensselaer as Stephen's "nephew," when he was a second cousin. W.W. Spooner, "The Van Rensselaer Family," *The American Historical Magazine*, Vol. 2, January 1907, No. 1, pp. 129-136.

(129) Originally called "Lewis Town," Lewiston was the Post-Township of Niagara County, 27.5 miles north from Buffalo. The British landed at Lewiston on December 19, 1813, and destroyed it along with other settlements in retaliation for the American destruction of Newark. It had been U.S. Army headquarters during the attack on Queenston a year earlier. Heidler, p. 301.

(130) This individual cannot be further identified from the information Myers provided. However, a Private Elihu Comstock of New London, Connecticut, was attached to Winfield Scott's 25th Brigade and was present at the Battles of Chippawa and Lundy's Lane. He died on March 1, 1815. Clarence Stewart Peterson, *Known Military Dead During the War of 1812* (Baltimore, Maryland: April 1955); Rept. Genealogical Publishing Co. Inc., 1991, p. 15. This may or may not be Myers's sergeant, as he does confuse names and ranks on occasion.

(131) New York native John Christie (1786-1813) entered the army in 1808 as a First Lieutenant in the *6th Infantry* and was promoted to Lt.Colonel of the *13th Infantry* on March 12, 1812. Christie believed that he, not militia Colonel Solomon Van Rensselaer, should command the attack at Queenston. Christie led 300 regulars across the Niagara River in the first wave of the assault on the night of October 12, arriving at Lewiston at midnight. Early the next morning he loaded his men into 13 boats, 3 of which were subsequently lost. Christie was on one of those 3, causing him to miss the fiercest fighting at Queenston, and the command went to one of his subordinates, Captain John Wool. Christie reappeared later that day during the American retreat, received a minor wound, was captured and exchanged later. Van Rensselaer claimed Christie was a coward but filed no charges against him. Christie was made Colonel of the *23rd Regiment* and participated in General Dearborn's campaign against Fort George in May of 1813. He died of natural causes at Lewiston on July 22, 1813, and was buried in the military cemetery at Fort Niagara. Chrystie Street in Manhattan's Lower East Side is named for him. Heitman, *HDRUSA*, vol. 1, p. 300; *HRUSA*, p. 61; Guernsey, vol. 1, pp. 97-98; Heidler, p. 105; Janet Carnochan, "Fort Niagara," *Niagara Historical Society:* No. 23 (orig. pub. 1922).

(132) William D. Lawrence of Rhode Island and New York served as a Captain of the *13th Infantry* beginning on March 12, 1812, and was honorably discharged June 15, 1815. Lawrence reenlisted on April 29, 1816, as battalion paymaster of artillery corps and was discharged again on November 20, 1817. He died February 26, 1833. Heitman, *HRUSA,* p. 402.

(133) Richard M. Malcolm of New York served as a Captain of the *13th Infantry* beginning April 8, 1812, and was wounded in the assault on Queenston Heights. He was promoted to Major on March 3, 1813, and Lieutenant Colonel on June 30, 1814. Malcolm was honorably discharged on June 15, 1815, and worked as a broker in Utica, NY. Heitman, *HRUSA,* p. 448; *Directory of Utica NY of Oneida County NY,* 1817, n.p.

(134) John Ellis Wool (1784-1869), a lawyer from Troy, NY, was captain in the *13th Infantry.* He led the charge up a fisherman's path at Queenston Heights which brought about the death of Isaac Brock (though in the end, the Americans lost). Wool went on to serve in the Mexican War and was the oldest serving General in the Civil War. Heidler, pp. 561-563.

(135) Henry Beekman Armstrong (1791-1884), the son of Secretary of War John Armstrong, served as a Captain in the *13th Infantry* beginning April 9, 1812. He was wounded during the Battle of Queenston, then promoted to Major of the *23rd Infantry* on April 12, 1813. Armstrong was present at the Battle of Stoney Creek, transferred to the *4th Rifles* March 25, 1814, and became Lieutenant Colonel of the *First Rifle Regiment* on September 17, 1814. He was honorably discharged June 15, 1815, and returned to his family's estate at Red Hook, Dutchess County, NY. At age 70 in 1861, Armstrong went to Washington and unsuccessfully offered his services to the Union Army. Heitman, *HRUSA,* p. 170; Lossing, p. 396, n. 3.

(136) Peter Ogilvie, Jr. of New York entered the army March 12, 1812, and served in the *13th Infantry* until resigning June 15, 1813. During the assault on Queenston Heights, Ogilvie and Wool captured an 18-pounder. Brigadier General Decius Wadsworth negotiated the release of both Ogilvie and Major James Mullany after they were captured during the battle. General Wool recommended, and Ogilvie received, a special commendation for bravery for his actions at Queenston, and later became a judge. Heitman, *HDRUSA,* vol. 1, p. 757; Lossing, p. 399, n. 2.

(137) There was no ballroom in Buffalo in 1812-13. Pomeroy's Tavern/ Hotel on Main and Seneca Streets was the likely location. It was a popular place for soldiers to congregate and where several officers boarded. Elijah D. Efner, "The Adventures and Enterprises of Elijah D. Efner: An Autobiographical Memoir." *Publications of the Buffalo Historical Society, Volume 4.* (Buffalo, New York: The Peter Paul Book Company, 1896), pp. 46-47.

(138) "McEwan Shoemaker" was directly across the street from Pomeroy's Tavern/Hotel. Buffalo's other 2 shoemakers were 3 blocks and too far from where many of the soldiers were boarding or camping in the street, so McEwan's may have been where Myers stayed. Frank H. Severance, "The Earliest Map of Buffalo including an 1813 Burned Buffalo Map," in Frank Severance, ed., *The Picture Book of Earlier Buffalo.* Buffalo Historical Society, Vol. 16, 1912, pp. 57-66.

(139) Myers's "old French fort" was Fort Schlosser, one mile above the Falls along the U.S. shore opposite Chippawa and garrisoned by the Americans until December, 1813, when it was captured and burned. That particular site had been established by the French as *Fort Petite Niagara* (Fort Little Niagara) to guard the upper landing of the primary portage around Niagara Falls, and was burned by its garrison to prevent use by British forces who were already assaulting Fort Niagara (some 15 miles to the north) in July 1759 during the French and Indian War. Fort Schlosser, Fort Little Niagara's successor, was built further east (upriver) in 1760 by the British. Still referred to as "Little Niagara," it was strengthened in 1763 by Captain Schlosser, in whose honor the garrison and builders dubbed it. It continued as "Little Niagara" on British maps and in private journals for a number of years afterwards, before "Schlosser" came into more common use. When the United States was given physical possession of the disputed Niagara territory in 1796 (which included Fort Niagara and the entire Niagara Portage on the U.S. side of the Niagara River) the name "Fort Schlosser" stuck.

(140) Native Virginian Samuel B. Archer (1791-1825) entered the army as an artillery corps captain on March 12, 1812 and was breveted Major in May 1813 for "gallantry and good conduct" during "the cannonade and bombardment of Ft. George." Archer served as a captain in Winfield Scott's Second Regiment of Artillery and was promoted to Inspector General with the rank of Colonel in 1821. Heitman, *HDRUSA*, vol. 1, p. 168; *HRUSA*, p. 92; Lossing, p. 602, n. 1.

(141) Hugh R. Martin (See Note #95).

(142) John Sproul (See Note #94).

(143) "Snake Island is close to Kingston and is so small that it does not show on most maps. During the War of 1812 there was a blockhouse on it, as it controls the channel entering Kingston Harbor. What Myers calls "rattle snakes" are actually water snakes fairly common in the area.

(144) James Crane Bronaugh (1788-1822) of Virginia was Surgeon of the *12th Infantry* on April 28, 1812. He was Hospital Surgeon on the General Staff of the Northern Army under Major General George Izard from April

15 to July 31, 1814, and Assistant Surgeon-General on April 18, 1818. Bronaugh was a close friend of Andrew Jackson's and became part of Jackson's "military family" during Jackson's governorship of Florida. He remained in Florida and died in Pensacola on September 2, 1822, during the August-September outbreak of yellow fever. Heitman, *HDRUSA*, vol. 1, p. 247; Andrew Jackson to General John Coffee, September 29, 1812, in Harold Moser, David R. Hoth and George H. Hoemann, eds., *The Papers of Andrew Jackson, 1821-1824* (Knoxville: The Unversity of Tennessee Press), Vol. V, p. 219; John Spencer Bassett and David Matteson Maydole, eds., *Correspondence of Andrew Jackson,* (Washington, D.C.: Carnegie Institution, 1926-1935), Vol. 2, Issue 371, p. 304, n. 8.

(145) John Stannard of Virginia enlisted as an Ensign in the *2nd Infantry* on December 9, 1807, was promoted to *2nd Lieutenant* October 8, 1809, and resigned September 1, 1810. Stonard was then commissioned as a Captain of the *20th Regiment* on March 12, 1812, and went on to become Major (March 3, 1813) and Lieutenant Colonel (June 28, 1814) of that regiment. He was honorably discharged June 15, 1815, and died September 23, 1833. Heitman, *HRUSA*, p. 59, p. 610; The Bronaugh-Stannard duel took place on May12-13, 1813. See Myers's *Reminiscenses*, pp. 54-55.

(146) Zachary Taylor (1784-1850) was the 12th President of the United States. He served as First Lieutenant of the *7th Infantry* (May 3, 1808), Captain and recruiter for the same infantry regiment (November 30, 1810), and Major of the *26th Infantry* on May 15, 1814. Taylor was in command at the Battle of Fort Harrison (September 4-5, 1812). After the war he served with the *3rd, 4th, 8th, 1st* and *6th Infantries* in New Orleans, Memphis, Louisville and Maryland. There is no evidence that Myers and Taylor served in any proximity to one another during the war, though Stannard, a Virginia native like Taylor, may have known the future President. Taylor was the senior Major with Colonel Isaac Clark when Clark commanded the defenses at Burlington at the head of Lake Champlain and was ordered to go to Plattsburgh; but, when peace was imminent he never served there. Heitman, *HRUSA*, p. 634; K. Jack Bauer, *Zachary Taylor: Soldier, Planter, Statesman of the Old Southwest* (Baton Rouge, Louisiana: Louisiana State University Press, 1985), p. 20.

(147) Thomas Beverly Randolph (d. 1867) was a Cadet at West Point (October 14, 1808), 2nd Lieutenant, then 1st Lieutenant of Light Artillery (January 3, 1812; January 20, 1813; in the interim participating at the Battle of Queenston), and Captain in the *20th Infantry,* April 15, 1813. Randolph resigned February 6, 1815, but became a Lt. Colonel of Hamtramck's *Regiment of First Virginia Volunteers* (or "Hamtramck's Guards") during the Mexican War, and participated in the capture of John Brown at Harper's Ferry. Heitman, *HDRUSA*, vol. 1, p. 815; *HRUSA*, p. 539.

(148) Myers mistakenly used "Snake Hill" instead of "Snake Island" in this instance. As for Snake Hill, when the Americans captured Fort Erie on July 3, 1814, they extended the earth wall to the south for an additional half mile to a 20-foot-high rise made of sand by Lake Erie, west of Old Fort Erie, called "Snake Hill." Myers, with army engineer Lt. Joseph Totten, constructed a gun battery there. Snake Hill also became known as "Towson's Battery." Paul Litt, Ronald F. Williamson, and Joseph W.A. Whitehorne, *Death at Snake Hill: Secrets from a War of 1812 Cemetery* (Ontario Heritage Foundation Local History Series No. 3. Toronto & Oxford: Dundurn Press, 1993), p. 32, pp. 83-85, p. 92, p. 95, p. 118, p. 135.

(149) Joseph Gilbert Totten (1788-1864) of Connecticut was a West Point graduate and Chief Engineer of the Niagara Frontier and Lake Champlain armies under General Stephen Van Rensselaer. Totten fought alongside Winfield Scott at Queenston, and when Scott was forced to surrender, it was Totten's cravat that Scott put to the end of his sword as a flag of truce. Totten was breveted Lieutenant Colonel for gallant conduct at the Battle of Plattsburgh. Heitman, *HDRUSA,* vol. 1, p. 966; Elting, p. 48; Lossing, p. 403, n. 1.

(150) James Trant (1750-1820), pronounced "Trent" as Myers wrote it, was an Irish-born sailing master of the U.S. Navy from its beginnings until the end of the War of 1812. Trant was master of the *Julia,* sailing from Sackett's Harbor on November 8, 1812, with Commodore Isaac Chauncey's flotilla. On November 9 *HMS Royal George* was being pursued by Chauncey; when the *George* reached the Canadian batteries at Kingston, Chauncey decided to follow to test the defenses and try to capture it. The *Julia* and *Conquest* followed until nightfall. On May 27, 1813, the *Julia* and *Growler* led the flotilla into the Niagara River and shelled the British battery near Fort George, allowing Commodore Perry to disembark troops. Fort George was taken in 3 hours and the British gave up their forts on the Niagara Frontier, which left the Americans in charge of the whole frontier. Trant, who slept with loaded pistols under his pillow, also participated in the assault on York, and was buried at sea at the time of his death. Edgar Stanton Maclay and Roy Campbell Smith, *A History of the United States Navy, from 1775 to 1893* (New York: D. Appleton and Company, 1894), Vol. 1, p.485; Lossing, p. 644, n. 1.

(151) Twelve Mile Creek is a waterway located in the Niagara Peninsula of Ontario, Canada, and flows through today's Thorold and St. Catharines, Ontario. It derives its name from the fact that its outlet to Lake Ontario is about 12 miles from the Niagara River.

(152) Pennsylvania native Patrick McDonough became a 1st Lieutenant of the *2nd Artillery* on March 12, 1812, and transferred to artillery corps May 12, 1814. He was killed August 15, 1815, in the defense of Fort Erie. Heitman, *HDRUSA,* vol. 1, 663; Hickey, 192.

(153) John Keyes Paige (1788-1857) graduated from Williams College and attended West Point before practicing law in Schenectady, NY, until March 17, 1812, when he joined the *13th Infantry*. There, he and Hugh R. Martin raised a company with members from throughout Schenectady County, and marched to Sackett's Harbor. Paige was an aide to General Leonard Covington, then General James Wilkinson. After the war, Paige served as Albany, New York's, District Attorney, Clerk of the NY State Supreme Court, Presidential Elector, Regent of the New York State University, and Mayor of Albany, NY. Heitman, *HDRUSA*, vol. 1, p. 765 ; Schenectady *Reflector*, December 18, 1857, p. 2.; Lucius Robinson Paige, *History of Hardwick, Massachusetts. With A Genealogical Register* (Boston: Houghton, Mifflin and Company, 1883), p. 447; *Hudson-Mohawk Geneological and Family Memoirs* (New York: Lewis Historical Publishing Company, 1911), Vol. II, pp. 665-666; "Diary of John Keyes Paige, 13th Regiment" in Ernest Cruickshank (ed.), *The Documentary History of the Campaign Upon the Niagara Frontier In the Year 1813. Part III (1813), August to October, 1813* (Welland, Ontario: The Lundy's Lane Historical Society), pp. 148-150.

(154) This individual was unable to be identified.

(155) Homer Virgil Milton (1781-1822), a native of Georgia, and son Georgia's first Secretary of State, began his military career as a Major in the *3rd Infantry* in 1808. By 1810 he was a Lieutenant Colonel in the *6th Infantry*, transferring to the *5th Infantry* in 1812. Milton was promoted to full Colonel in the *3rd Infantry* in 1813, and resigned in November, 1814. Heitman, *HDRUSA*, vol. 1, p. 714.

(156) On May 27, 1813, a division under Colonel James Burns crossed at Five Mile Meadows (located 5miles upriver from Fort Niagara, and the site of boatbuilding from Fall 1812 until the attack on Fort George, May 1813) and took Fort Erie with no resistance. Burns, a native of South Carolina who later resided in Virginia and Pennsylvania, had been a cavalry officer in March 1799 during the "Quasi-War" with France. He was commissioned as a Colonel and commander of the *Second Light Infantry Dragoon Regiment* on July 6, 1812. Burns' job on May 25 was to prevent the British retreating from Fort George to reach Queenston, but the British batteries prevented them from doing so. General Dearborn called off the attack, which allowed the entire British force to escape to Burlington Heights. This led to the Battles of Stoney Creek (June 5) and Beaver Dams (June 24). Myers was incorrect about Burns taking Chippawa. The Battle of Chippawa took place on July 5, 1814. Myers may have confused this with the American capture of Fort George on May 27, 1813. Heitman, *HDRUSA*, vol. 1, pp. 204; *HRUSA*, p. 38; Heidler, pp. 70-71.

(157) General Morgan Lewis (see Note #32) was in pursuit of the British (June 1-4, 1813), who retreated to Burlington Heights.

(158) James Crooks (1778-1860) was a Scottish-born businessman, politician, Justice of the Peace and militia officer who came to Newark (Niagara-on-the-Lake) in 1791, and established Crook's Mill on the Trent River by 1810. Crooks commanded a flank company at Queenston under General Sheaffe. In December, 1813, the Americans seized and destroyed "Crookston" (or "Crooks' Hollow"), Crooks' homestead that included outbuildings and storehouses. *DCBO*, VIII. 1851-1860.

(159) The Cook family had been settled on the west bank of the Niagara River since 1785. *Niagara Falls: A Chronicle of Our Early Settlers – A History, 1600-1900*, p. 11, p. 30. http://www.niagarafrontier.com/work.html. Hereafter cited as *NF.*

(160) A brigade under General William H. Winder first followed Vincent, but he decided that Vincent's forces were too strong to engage. Winder halted at Forty Mile Creek, where they were joined by General Chandler's brigade before advancing to Stoney Creek.

(161) Commodore Sir James Yeo (1782-1818) joined the British navy as a Midshipman at age 10 and was a veteran of the Napoleonic wars. He was sent to Canada to command British naval forces on the Great Lakes, but Sir Geroge Prevost failed to support or follow through on Yeo's advances at Sackett's Harbor or elsewhere. But, Yeo's efforts ensured control of Lake Ontario. He had several commands in West Africa and the Caribbean, and died while returning to England from Jamaica. *DCBO*, Vol. V, 1801-1820; Heidler, pp. 567-568.

(162) Samuel B. Archer of Virginia was a Captain in the *Second Artillery* under Brigadier General Moses Porter, March, 1812. He was transferred to the artillery corps May 12, 1814, and to the *Third Artillery* June 1, 1821. Archer was promoted to Colonel I.G. November 10, 1821 and was breveted Major on April 27, 1813, for gallantry and good conduct in the cannonade and bombardment of Fort George. Heitman, *HDRUSA*, vol. 1, p. 168; Graves, *FOG*, p. 358.

(163) This is probably David Vanderheyden of New York, a 2nd Lieutenant of the *6th Infantry*, June 3, 1812, and 1st Lieutenant on February 21, 1814. The *6th Infantry* participated in the Battle of Stoney Creek under Lt. Colonel James Miller. Heitman, *HDRUSA*, vol. 1, p. 981; Graves, *FOG*, p. 356; James E. Elliott and Nicko Elliott, *Strange Fatality: The Battle of Stoney Creek, 1813*. (Montreal, Quebec, Canada: Robin Brass Studio, 2009), p. 28, p. 39, p. 43, p. 160, p. 171, p. 187. Hereafter cited as *Elliott.*

(164) John Chandler (1762-1841) had been an illiterate, self-educated blacksmith who served in the *New Hampshire State Militia* during the Revolution. With financial help from Henry Dearborn he became wealthy, settled on a farm in Monmouth, Maine (then part of Massachusetts), and went on to serve in the Massachusetts State Senate, U.S. House of Representatives, and as sheriff of Kenebec County. Chandler was wounded and captured at Stoney Creek after wandering into the British lines. After the war he was a member of the Massachusetts General Court, first president of the Maine Senate, a U.S. Senator, trustee of Bowdoin College, Major-General of the State Militia and Collector of the Port of Maryland. Heitman, *HDRUSA,* vol. 1, 295; vol. 2, 613-14; Lossing, p. 603, n.2; Andrew R.Dodge, *Biographical Dictionary of the United States Congress 1774-2005* (Washington, D.C.: U.S. Government Printing Office, 2006), p .806. Hereafter cited as *Dodge.*

(165) Myers refers here to General John Vincent (1765-1848), who had disappeared during the Battle of Stoney Creek (June 6, 1813). Vincent was still missing for much of that morning, having been thrown from his horse and injured during the battle, as well as losing his horse, hat and sword. He became lost in the woods and hiked west for 7 miles until finding the British lines. Elting, p. 127; Mahon, p. 151; Elliott, et passim.

(166) Myers refers here to General Vincent, but the fictional character Myers describes cannot be identified.

(167) For the most comprehensive account to date on the Battle of Stoney Creek, see Elliott.

(168) Major-General Francis Baron de Rottenburg (1757-1832) was a veteran of the French army during the French Revolution. He received a commission in the British army during the Napoleonic wars and "was sent to Canada in 1810 as a brigadier general and was promoted the following year to major general." (Heidler, pp. 144-45). Rottenburg replaced Major-General Roger Halc Sheaffe as Chief Military and Civil Commander of Upper Canada on June 19, 1813. Sheaffe had replaced General Sir Isaac Brock, who had been killed at Queenston Heights.

(169) As one historian has pointed out, "Myers solution to the night-time sniping was simple but effective." Graves, *FOG,* p. 38.

(170) During the "Skirmish at Butler's Farm" (July 8, 1813), a detachment of Native warriors, some provincial dragoons, and men from the *8th Regiment* were trying to recover medical supplies at Two Mile Creek near Fort George when they encountered American pickets, including 40 soldiers of the *13th Infantry* led by Lt. Joseph C. Eldridge. Native warriors concealed in a ravine ambushed them, killing and mutilating 28 Americans;

12 others were taken prisoner. Eldridge Street in Manhattan's Lower East Side was named for Lt. Eldridge in 1817. Graves, *FOG,* pp. 38-41; Heitman, *HRUSA,* p. 253; Lossing, p. 626, n. 3, n. 4.

(171) This was the site of the Battle of Lundy's Lane, July 25, 1814, which has been called "one of the bloodiest battles of the war on Canadian soil." See Donald E. Graves, *Where Right and Glory Lead! The Battle of Lundy's Lane, 1814* (Toronto, ON, Canada: Robin Brass Studio, Incorporated, 1997).

(172) Charles Wilson and his wife were the first innkeepers at the Falls and opened a hostelry, "The Exchange," in Newark on the River Road. Their log hut, renamed "Wilson's Tavern" (also spelled "Willson's Tavern"), appears to have been a meeting place and informational clearinghouse of sorts for both sides during the war. Richard Merritt, Nancy Butler, and Michael Power, eds., The *Capital Years: Niagara-on-the-Lake, 1792-1796* (Toronto, ONT: Dundurn Press Limited, Co-published by the Niagara Historical Society, 1991), p. 216; *NF,* p. 14.

(173) St. David's, Ontario, Canada, is a village included in the Town of Niagara-on-the-Lake (formerly Newark). On July 18, 1814, Major General Peter B. Porter sent a militia detachment from Queenston to attack the village. Regulars under Lt. Colonel Isaac Stone joined them in looting and burning most of the village's 30-40 houses. Stone was censured and dismissed from the U.S. Army for this action. Lossing, p. 815, n. 1.

(174) Columbia County (NY) attorney John Van Hoesen Huyck was a Major in the *23rd Infantry* in April, 1812, and transferred to the *13th Infantry* on August 26, 1812. He was honorably discharged June 15, 1815, and died February 10, 1829. Heitman, *HDRUSA,* vol. 1, p. 561; Franklin Ellis, *History of Columbia County, New York* (Philadelphia: Everts & Ensign, 1878), p. 112.

(175) Robert Hamilton (1753-1809) was a Scottish-born graduate of the University of Glasgow who came to Canada with a fur company west of the Great Lakes and settled in Queenston in 1784. He formed the first organized portage between Chippawa and Queenston along the Portage road in 1788 with various partners. Hamilton established military contracts to supply goods to the British army at Fort Niagara starting in 1791 and went on to serve Judge of the Court of Common Pleas and member of the Legislative Council for the Province. Hamilton amassed sufficient wealth through land speculation to build an impressive Georgian mansion on the Niagara Escarpment at Queenston that boasted a 2-story façade, but the house was damaged during the war and his family was left homeless. *DCBO,* V, 1801-1820; Bill Freeman, *Hamilton: A People's History* (Toronto, Ontario: James Lorimer & Company Ltd., 2001), pp. 26-27.

(176) Major Cyrenius Chapin (1764-1838), a Buffalo physician and surgeon, secured a Major's position and organized a unit of mounted New York Militia volunteers known as "Dr. Chapin and the Forty Thieves" because of their looting raids across the Niagara River. In late June 1813, Chapin helped convince American officials at Fort George to attack a British outpost 16 miles away with a force that included 600 troops under Lieutenant Colonel Charles Boerstler, and 38 volunteers under Chapin. Boerstler, technically (as of June 20th), held the rank of Colonel when the plan was put into action on June 23rd, but was not yet aware that his promotion had been approved in Washington. Meanwhile, British Lieutenant James Fitzgibbon learned of the advancing force from Native scouts, and critical details of the planned attack from Canadian civilian Laura Secord who overheard several U.S. officers with loose lips (see also Note 180). On June 24th, at the Battle of Beaver Dams, British Indian Department forces attacked the American column, and Fitzgibbons — in a masterful stroke of guile — was able to capitalize on the resulting confusion to use false statements of tactical superiority and implied threats of unrestrained Native brutality and to bluff Boerstler into surrendering his entire force without British regulars having to fire a shot. Chapin and some of his men were made prisoners but escaped while being taken to Kingston, and returned to Fort George. Hickey, p. 138; Lossing, p. 622.

(177) William McGillivray (1764-1825) served in the Legislative Assembly of Lower Canada for Montreal West. As a Lieutenant Colonel in the British army, he commanded the "Corps of Canadian Voyageurs", a battalion-strength militia who participated in two engagements. McGillivray was taken prisoner by Captain Simeon Wright while scouting the Canadian line in 1814. *DCBO*, VI, 1821-1835; Robert McGillivray and George B. MacGillivray, *A History of the Clan MacGillivray* (privately printed, 1973), p. 21; Heidler, p. 26, p. 44. Simeon Wright of Vermont was a captain in the *30th Infantry Regiment.* See Heitman, *HDRUSA,* vol. 1, p. 1063.

(178) The vedette (Myers's "vidette") described here would have been a mounted scouting party moving ahead of the main force in order to observe, spot, scout, etc., for anything having to do with the enemy.

(179) The identity of this surgeon cannot be determined conclusively, based on the information provided by Myers and existing records.

(180) Myers "Butler" was actually Charles G. Boerstler (1778-1817), the son of Dr. Christian Boerstler. Charles arrived in America from Bavaria, Germany, in 1784, and entered the army as a Lieutenant Colonel on March 12, 1812. He marched 300 men to join General Smyth on the Niagara Frontier where he commanded the *14th Regiment,* which spearheaded the failed invasion of Canada. On June 23, 1813, Boerstler's force of regurs,

along with Chapin's Volunteers, moved from Fort George to Queenston and housed themselves in private homes, including that of James and Laura Secord. James Secord had been severely wounded at Queenston the previous year and had no love for the Americans, so when Laura overheard plans for the American attack, she walked 20 miles through the woods to report the information to the British. Boerstler surrender at Beaver Dams the next day, June 24th, and along with Brigadier General William H. Winder was made a prisoner of war pending parole. A court of inquiry found Boerstler's surrender justified and he continued to serve until honorably discharged on May 17, 1815. He died at New Orleans November 21, 1817. As discussed in Note 176, he was technically a Colonel at the time of the Battle of Beaver Dams, but was not aware that his promotion to that rank from that of Lieutenant Colonel had been approved, effective June 20th. Heitman, *HRUSA*, p. 54; *HDRUSA*, vol. 1, p. 108, p. 227; vol. 2, p. 15; Heidler, 58-59; John Thomas Scharf, *History of Western Maryland, Being a History of Frederick, Montgomery, Carroll, Washington, Allegany, and Garrett Counties* (Philadelphia: L.H. Everts, 1882), p. 1047; Dr. Christian Boerstler, "Autobiography of Bavarian Immigrant: 'I Was a Liberty-Loving Man'." *The Connecticut Magazine*, Vol. XII, Spring of 1908, Number 1, pp. 401-414.

(181) William Henry Harrison (1773-1841) led U.S. forces against Native Americans at Tippecanoe (November 7, 1811) and was commanding general at the Battle of the Thames (also known as the Battle of Moraviatown) on October 5, 1813, which resulted in the death of the Shawnee Chief Tecumseh and the destruction of the Native coalition he led. Harrison became the 9th President of the United States and the first to die in office. Heidler, pp. 230-232.

(182) James Wilkinson (1757-1825) served on General Gates' staff at the Battle of Saratoga during the Revolutionary War and was later involved in the "Conway Cabal" plot against George Washington. Later, still, he was "Clothier-General to the army" but resigned due to irregularities in his accounts. Wilkinson was the Federal Commander of the Gulf District in the Southwest when he became involved in the Burr Conspiracy. He received a pension from the Spanish government as "Spy No. 13" and faced a Congressional court of inquiry and court-martial, but cleared himself by testifying against Burr, and was found not guilty. He went on to command the American forces during the unsuccessful St. Lawrence campaign. *DAB*, X, 222-226; Heidler, 553-555; For a readable biography of Wilkinson, see Royal Ornan Shreve, *The Finished Scoundrel* (Indianapolis: The Bobbs-Merrill Company, 1933).

(183) Wade Hampton (1752-1835), one of the largest landowners and slaveholders in South Carolina and Mississippi, served 2 terms in Congress and received a Colonel's commission in the Army in 1808. He was promoted to Brigadier General in 1809 and appointed by James Madison to oversee the defenses at Norfolk, VA. On July 3, 1813, Hampton was given command of a division at Burlington on Lake Ontario, under General Dearborn. Graves, *FOG*, pp. 31-32; *DAB*, IV, pp. 212-213; Heidler, pp. 226-227.

(184) Chateaugay, NY, bordered by Quebec, Clinton and Essex Counties, Malone and Constable, is also referred to as "Chateaugay Four Corners" after the 4 lots of the township corner located on the Old Military Tract granted to Charles LeMoyne and later settled by the area's leading families. Colonels John Wool and Josiah Snelling were stationed there for a time. Chateaugay had two blockhouses during the War of 1812. General Wade Hampton arrived there with several thousand men to support Wilkinson's move against Montreal, but after a skirmish there he withdrew to Plattsburgh. Frederick J. Seaver, "History of Chateaugay, New York," in Seaver, Chapter XI.

(185) "McClear" was Irish-born Brigadier General George McClure (1771-1851), an attorney who served as a New York State Assemblyman, Sheriff, Surrogate Judge and County Judge in Bath, NY, and was owner-operator of a gristmill and ferry on Cayuga Lake. He had been left in command of Fort George as the St. Lawrence campaign was ending and destroyed or burned more than 80 civilian buildings in Newark on December 10th, 1813, leaving mainly women and children (whose husbands and fathers were in the service or were American prisoners) to fend for themselves in sub-zero temperatures. McClure claimed he had been ordered to burn the town. Both sides condemned the action and he was eventually relieved of his command and dismissed from the Army. After burning Newark and evacuating Fort George, he went to Buffalo and left General Amos Hall in charge of Fort Niagara. Colonel John Murray took the fort in the early morning of December 19th. Babcock, 115-130; Hickey, 142.

(186) Vermont native Nathaniel Leonard began his military career as a Lieutenant in the newly-formed *Second Regiment of Artillerists and Engineers* (June 1, 1798), and then placed in the *First U.S. Regiment of Artillerists* (April 1802) when it was created by the 1802 separation of artillerists and engineers into distinct corps; he was promoted to Captain on December 1, 1804. Further restructuring was authorized in January of 1812, and Leonard was made a Company commander in the *1st Regiment of Artillery* (March 12, 1812; known as *Captain Nathaniel Leonard's Company*). At the time of Fort Niagara's capture in December of 1813, the garrison there was com-

posed of Leonard's company of artillerymen, one company of soldiers from the *24th U.S. Infantry* (Tennesseeans), and small detachments of wounded and sick from other units. Murray's night assault on Fort Niagara by British infantry led to Leonard's capture at his home 2 miles away, where he was found intoxicated. Despite his reputation as a heavy drinker and poor soldier who lost Fort Niagara, he was honorably discharged on June 1, 1814. Heitman, *HDRUSA,* vol. 1, 628; Lossing, 631; Elting, 152-54.

(187) In a 4-day period (December 18-19, 29-30, 1813), British troops under Colonel John Murray and Major Phineas Riall also destroyed Manchester (Niagara Falls). They also attacked Fort Schlosser, where many of the troops there were mutilated, dismembered and beheaded in retaliation for McClure's destruction of Newark on December 10th (1813). Babcock, 115-130; Quimby, vol. 1, 353-360; Hickey, 142-43; Hitsman, 194-95.

(188) Mexico is a town located in the northeast part of present-day Oswego County, New York; it contains a village, also named Mexico. The original settlement of Mexico was created in 1792 in what was then the Town of Whitestone (when that entity was located in the Herkimer County of the time). The Town of Whitestone was re-established in 1796, expanded, then reduced, as new Towns were formed and incorporated from its vast expanse of land.

(189) Joseph Willcocks (1773-1814) was a diarist, officeholder, printer, publisher, journalist, politician and army officer who is considered a traitor in the annals of Upper Canada. He was disturbed when, after the invasion of the Niagara Frontier in 1813 (particularly the capture of York and the British victory at Stoney Creek), there was the imposition of harsh military measures against anyone considered disloyal. At the end of August 1813 he formed a unit of expatriate Upper Canadians known as the "Company of Canadian Volunteers" (mainly recent immigrants from the U.S.) while he was a sitting member of the Legislative Assembly of Upper Canada. Willcocks was shot in the chest and died while leading skirmishes during the September 1814 Siege of Fort Erie. *DCBO,* V, 1-5; Donald E. Graves, "Lawless banditti: Joseph Willcocks, his Canadian Volunteers and the Mutual Destruction on the Niagara during the Winter of 1813" [Part I], *Fortress Niagara* (Journal of the Old Fort Niagara Association): Vol. IX, No. 2, December 2007, pp. 4-9; and [Part II (Conclusion)] Vol. IX, No.3, March 2008, pp. 4-9.

(190) John Morton of New York was a Captain in the Department of Commissary and Ordnance on September 11, 1812. He was retained as a Captain of Ordnance on February 8, 1815, and discharge June 1, 1821. Heitman, *HDRUSA,* vol. 1, p. 731; *HRUSA,* p. 480.

(191) Grenadier Island, near the head of the St. Lawrence, consists of 1290 acres, measures 2.3 x 1.4 miles, and had a watch post on it during the war. The island was first noted by the Jesuit priest Pere Simon le Moyne in 1654; Champlain, LaSalle and Frontenac stopped there temporarily. The Mississaugas, who called it "Toniata" ("beyond the point") controlled it until United Empire Loyalists settled there in the early 19th-century. During the Embargo period it was used by smugglers as a rendezvous. It was "the final staging point for the attack on Kingston" Graves, *FOG*, p. 61; Edgar C. Emerson, ed., *History of Cape Vincent,NY* (The Boston History Company, Publishers: 1898), n.p.; James White, *Place-Names In The Thousand Islands, St Lawrence River* (Ottawa: Government Printing Bureau, 1910), p. 5.

(192) John Armstrong, born November 25, 1758, in Carlisle, Pennsylvania, had been a student at Princeton when he served on General Gate's staff in the Revolution. He was the author of the "Newburgh Addresses," served as Secretary of State and Adjutant-General of Pennsylvania, married the sister of New York State Chancellor Robert R. Livingston, and farmed within the old Livingston Manor at Red Hook on the Hudson River. Armstrong became a U.S. Senator, and U.S. Minister to France as successor to his brother-in-law Chancellor Livingston, and was commissioned Brigadier General in July 1812. In January 1813 he became Secretary of War in James Madison's cabinet, but resigned because he failed in his operations against Canada and in his efforts to defend Washington in 1814. Heidler, pp. 13-16.

(193) There was no known physician or surgeon named McNear at that time or place, but Myers may very well be referring to merchant-shipper Matthew McNair of Oswego, New York. McNair, born in Scotland in 1773, settled in Oswego in 1802, acquired a schooner, and engaged in forwarding freight over Lake Ontario. He purchased additional Canadian vessels before building his own. By 1806, in conjunction with Jonathan Walton & Co. of Schenectady and other firms, he operated most of the transit business on the Great Lakes along the route from Albany and Schenectady to Wood Creek, then Oswego, then on to Lake Ontario, Lewiston on the lower Niagara, and then to Black Rock on the upper Niagara near Lake Erie. The U.S. government bought and armed all available craft in the region, the first being the *Julia,* built for cargo by McNair but now manned by riflemen and armed with a 32-pounder and two 6-pounders. McNair's schooner *Diana* was also acquired for Isaac Chauncey's Lake Ontario Squadron. The outbreak of war ended commercial transactions on the lakes, but McNair was appointed the U.S. Commissary of Subsistence at Oswego. He also participated directly in the war; on May 5, 1814, he and several others occupied the "Half Moon Battery" at Oswego's old French fort in anticipation of an attack. McNair continued to

be one of Oswego's most prominent citizens, serving as its "President" (mayor), a founder of its first Masonic Lodge, and one of the Town's Supervisors. McNair had a personal and commercial investment in Oswego, and if he did indeed zealously participate in the rescue by expertly dealing with the sick, wounded and dead. Myers may well have mistaken him for a surgeon. *History of the Great Lakes* (Chicago: J.H. Beers & Co., 1899), Vol. I, p. 131; John B. McLean, *A Brief History of the First Presbyterian Church of Oswego, N.Y. From Its Organization, November 21, 1816 to the Present* (Oswego: Times Book and Job Printers, 1890), p. 47; R. Pearsall Smith, *Gazetteer of the State of New York* (Syracuse, N.Y.: J.H. French, 1860), p. 525; J. Gerald Stone, *1895 Landmarks of Oswego County, NY* (September, 2004), pp. 131-132.

(194) John Parker Boyd (1764-1830) joined the regular army in 1784 as an Ensign, then was promoted to Lieutenant until 1789. He resigned and hired himself out as a mercenary in India for 19 years. When he returned, he obtained a Colonel's commission and command of the *4th Infantry,* which he led into action under William Henry Harrison at Prophetstown and Tippecanoe. Boyd was promoted to Brigadier General in 1812 and served under Generals Dearborn and Wilkinson on the Niagara Frontier, including the taking of Fort George in May, 1813. Later that year, Wilkinson ordered Boyd to clear out the British land force harassing the American flotilla as it descended the St. Lawrence, but he was unorganized and failed in his assignment. The result was the British victory over the Americans at Crysler's Field, November 11, 1813. Heidler, pp. 60-61.

(195) Durham boats were large wooden vessels produced by the Durham Boat Company of Durham, Pennsylvania, starting in 1750. They were designed to navigate the Delaware River and transport products made by the Durham forges and Durham Mills to Trenton, New Jersey and Philadelphia, Pennsylvania. These large, flat-bottomed boats were 60 feet long and 42 inches wide; both ends were tapered so either could serve as the bow. Its heavy steering gear could be shifted, along with setting poles or oars. A Durham boat could carry 17 tons downstream and 2 tons upstream, plus steersmen to operate it. George Washington's crossing of the Delaware included the use of Durham boats, and they were employed by both sides during the War of 1812. "History of the Durham Boat," *The Durham Township Historical Society Online,* 20 January 2012 (http://durhamhistoricalsociety.org/history2html).

(196) William Morrey Ross of New York city was a Surgeon's Mate with the 23rd Infantry on July 6, 1812, and hospital surgeon March 18, 1813. He was honorably discharged June 15, 1815. Heitman, *HDRUSA,* vol. 1, p. 847; *HRUSA,* p. 561.

(197) James Patton Preston (1774-1843) was a Virginia planter who attended the College of William and Mary and served 4 years in the Virginia State Senate and two years in the Virginia House of Delegates. He became a Lieutenant Colonel of the *12th Infantry* in 1812, then Colonel of the *23rd Infantry Regiment* from August 15, 1813, to May 17, 1815. Preston was "maimed for life" at Crysler's Field (November 11, 1813) but went on to serve again in the Virginia House of Delegates, and was elected the 20th Governor of Virginia for 3 one-year terms (1816-1819). He was Postmaster at Richmond, Virginia, until he retired. Preston County, Virginia, was named for him in 1818. Heitman, *HDRUSA*, vol.1, p. 121; William Allen, *The American Biographical Dictionary* (Boston: John P. Jewett and Company, 3rd ed., 1857), p. 679.

(198) The *USS Lady of the Lake* was an 89-ton vessel with a 15-man crew and one 9-pounder gun, built by Henry Eckford at Sackett's Harbor sometime between summer and winter of 1812-13. It was launched April 6, 1813, and entered service April 19th under the supervision of Commodore Isaac Chauncey as a dispatch boat on Lake Ontario, carrying messages to Niagara. It was used during the assault on York and the attack on Fort George, and engaged the British under Yeo on September 11, 1813, on Lake Ontario. The *Lady of the Lake*, assisted by three American ships, helped to capture several British schooners on October 5, 1813. "USS Lady of the Lake," *Dictionary of American Naval Fighting Ships* (Naval History & Heritage Command: Department of the Navy – Naval Historical Center, Washington, D.C), www.history.navy.mil/danfs/I1/lady_of_the_lake-i.htm.

(199) Joseph Wanton Morrison was born in New York in 1783, "son of the deputy commissary general of British North America." He first served in 1799 with the *17th Foot* in Holland, then garrison and staff duty in Minorca, Ireland, England and Guernsey. By 1809 he was Major of the *89th Foot*, then saw promotion to Lieutenant-Colonel, commanding the regiment in Trinidad, then Great Britain. In May 1812 he was in command of the Second Battalion of the *89th Foot*, and transferred with his unit to North America. Morrison was wounded at Lundy's Lane, and after the war was honored for his victory at Crysler's Field. He was promoted to Colonel and sent to India, but died of malaria in 1826. Morrison was buried at sea. Graves, *FOG*, pp. 134-35, pp. 330-31, p. 362.

(200) Ogdensburg, New York, was the scene of several military operations, including the Battle of Ogdensburg of February 21-22, 1813. The British gained a victory over the Americans, captured the village and removed the American threat to British supply lines for the rest of the war. Lossing, 177.

(201) Prescott is a town on the north shore of the St. Lawrence River in Ontario, Canada, founded by Loyalist Edward Jessup.

(202) A "barbette" is a protective circular armor feature around a cannon or piece of heavy artillery.

(203) Early on Monday, November 8, 1813, General Wilkinson moved his flotilla "and it dropped down to a bay in front of the 'White House' (near modern Sparrowhawk Point) about three miles below Tucker Brook on the American bank", situated "where the river narrowed." Graves, *FOG,* pp. 142-43.

(204) The Long Saulte (various spellings), 9 miles in length, was part of a series of rapids now called "the International Rapids" that Wilkinson's flotilla would have to run to reach Montreal. Graves, *FOG,* p. 116; Quimby, II, p. 340; For a vivid description of the running of the Long Saut about 60 years prior to the War of 1812, see Thomas Hamilton, *Men and Manners in America* (Edinburgh: William Blackwood, 1833; Rpt. New York: Augustus M. Kelley, Publishers, 1968), pp. 378-379.

(205) Cornwall is a city in Eastern Ontario, Canada, on the St. Lawrence River. It was originally settled by native/aboriginal peoples, then by United Empire Loyalists from New York.

(206) John Crysler's farm, where the battle took place on November 11, 1813, was "a few miles from the Canadian hamlet of Wiliamsburg." Donald E. Graves, *Soldiers of 1814: American Enlisted Men's Memoirs of the Niagara Campaign* (Youngstown, New York: Old Fort Niagara Association, Inc., 1995), p. .27, n.39. For the definitive work on the Battle of Crysler's Field, see Graves, *FOG,* et passim.

(207) Robert Swartwout (1778-1848; incorrectly spelled as Swartout or Swarthout at times) was a wealthy New York merchant who was commissioned a Colonel in the New York State Militia in 1812 and Brigadier General and Quartermaster-General in March, 1813. After his honorable discharge in 1816, Swartwout returned to mercantile pursuits and was a naval agent in New York. Graves, p. 7 (reference to the article "Hard School of War...", note 99)

(208) Major Richard M. Malcolm (see note #133) commanded the Boat Guard of 600 men under Lt. Col.Timothy Upham during the descent down the St. Lawrence. Graves, *FOG,* p. 361.

(209) Timothy Upham (1783-1855; other sources state he died 1851) was a storeowner in Portsmouth, New Hampshire, in 1807. He was commissioned a Major of U.S. Infantry on March 12, 1812, went to Plattsburgh, New York, in September, and then on to recruiting duty in Maine. As a Lieutenant Colonel of the *21st Infantry* (he had been promoted before the Battle of Crysler's Field) he rescued Colonel James Miller from Fort Erie (Septem-

ber 14, 1814) on the order of General Jacob Brown. Upham was suffering from poor health when he was honorably discharged June 15, 1815. After the war he became Collector of Customs at Portsmouth, Major General of the First Brigade, First Division of the New Hampshire Militia, engaged in commercial pursuits, ran unsuccessfully for Governor and retired after serving as the Naval Agent at Portsmouth. Heitman, *HRUSA*, p. 656; James Grant Wilson and John Fiske, *Appleton's Cyclopedia of American Biography*. (New York: D. Appleton and Co., 1889), Vol. VI, p. 212.

(210) Maryland native Ninian Pinkney entered the army as a First Lieutenant of the *9th Infantry* in 1799, became Regimental Paymaster in March, 1799, and was honorably discharge June 15, 1800. Upon re-entering the service he rose in the ranks to Major (January 1, 1813), Lieutenant Colonel of the *22nd Infantry* (April 15, 1814-May 17, 1815), Colonel in the Inspector General's Department (December 1, 1813-June 15, 1815) but returned as a Lieutenant Colonel of the *2nd Infantry* (May 17, 1815). Pinkney became Colonel of the *6th Infantry* on May 13, 1820, and died in 1825. Heitman, *HDRUSA*, vol. 1, p. 793.

(211) The *13th Infantry* was known as "The Jolly Snorters" because it was their style to wear extravagant mustaches as ordered by Lieutenant Colonel John Chrystie. On the march to the Niagara Frontier in 1812, some of the officers conspired to get everyone in the regiment to shave off their mustaches when Chrystie forbade them permission to visit a local town. Robert G. Malcomson, *Historical Dictionary of the War of 1812*. (Lanham, MD: Scarecrow Press, 2006), p. 554. According to Mordecai Myers, the "local town" was Utica, New York. See Myers's *Reminiscenses*, p. 49.

(212) Jonas Cutting, Jr., served in the American Revolution when quite young while growing up in Massachusetts and Vermont. He became a Lieutenant Colonel of the *25th Infantry* on March 12, 1812, and resigned from the Army on September 1, 1814, moving to Weathersfield, Vermont, where he owned a farm. He died on either April or August 5, 1834, at Woodstock, Vermont. Heitman, *HDRUSA*, vol. 1, p. 124, p. 349; *HRUSA*, p. 62; *American Ancestry: Giving the Name and Descent, in the Male Line of Americans Whose Ancestors Settled in the United States Previous to the Declaration of Independence, A.D. 1776* (Albany, N.Y.: Joel Munsell's Sons, 1894), Volume 9, p. 22.

(213) William Wallace Smith of New Jersey was an 1809 graduate of West Point. He became a Second Lieutenant in the *Light Artillery* (June 1, 1812), then First Lieutenant of *Light Artillery* on October 11, 1813. Smith was engaged in the capture of Fort George and the defense of its outposts, and was mortally wounded at Crysler's Field while manning a piece of artillery, himself, after all of his men were killed. Smith died a prisoner at Fort

Prescott on December 13, 1813. Heitman, vol.1, p .905; Cullum, vol. 1, p. 109; Lossing, p. 653, n.3.

(214) William Jenkins Worth (1794-1849) of Hudson, New York, had been a storekeeper prior to entering the army in May 1813 as a Lieutenant of Infantry, and later distinguished himself at the Battles of Chippawa and Lundy's Lane, where he was severely wounded. He was in commandant of cadets at West Point (1820-28), Colonel of the *8th U.S. Infantry* (1838), participated in the Seminole War (1842) and was in command of the army at Florida (1841-42). Worth was breveted a Brigadier General in March 1842 for service in the "war against the Florida Indians," and commanded a brigade under General Taylor in Mexico in 1846, distinguishing himself in the capture of Monterey, for which he was later breveted a Major General. In 1847-48 he commanded a division under Winfield Scott in the capture of Vera Cruz, Cerre Gordo and Mexico City. Worth was presented a sword by the United States Congress and further honored by Columbia County, New York, and the New York State Assembly. He died of cholera at San Antonio, Texas, May 17, 1849. His remains were returned to New York for burial, and re-interred with great ceremony on November 25, 1857, in the newly constructed Worth Monument in New York City (Worth Square, Manhattan). Fort Worth, Texas, was named for him. Donald E. Graves, (ed.), *First Campaign of an A.D.C.:The War of 1812 Memoir of Lieutenant William Jenkins Worth, United States Army* (Youngstown, New York: Old Fort Niagara Association, 2012); Heitman, *HDRUSA*, vol. 1, p. 1061; *HRUSA*, 713; Heidler, pp. 563-564; *DAB*, X, pp. 536-37.

(215) William Anderson, Jr., of Pennsylvania entered the army as a 3rd Lieutenant in the *13th Infantry* on May 13, 1813, and was promoted to 2nd Lieutenant (August 15, 1813), then 1st Lieutenant (October 1, 1814) before being honorably discharged June 15, 1815. He was reinstated as a 1st Lieutenant of Ordnance on May 17, 1816, and dropped September 1, 1818. Anderson died September 13, 1839. Heitman, *HDRUSA*, vol. 1, p. 165; *HRUSA*, p. 89. His father, Irish-born William Anderson, fought in the American Revolution and later became a construction contractor who built the first "Executive Mansion" in Washington. The elder Anderson died in 1821. John W. Jordan, (ed.), *Colonial and Revolutionary Families of Pennsylvania: Genealogical and Personal Memoirs* (New York & Chicago, 1911), Vol. I, p. 897. Reprinted (Baltimore, Md.: Clearfield Co. Inc., by Genealogical Publishing Co., Inc., 1994, 2004).

(216) The "complete victory" that day at Crysler's Field belonged to the British and Canadian forces and their Native allies. The total strength of the British forces was about 1169, while for the Americans it has been estimated at 3050. Graves, *FOG*, 361, 363.

(217) Henry Atkinson (1782-1842) of North Carolina entered the army in 1808 as Captain of the *3rd Infantry* and rose in the ranks to Colonel in the Inspector General's Department by April 25, 1813. Over the years he was promoted to Brigadier General, established Fort Atkinson at Council Bluffs, Nebraska, and as Indian Commissioner, led 476 troops up the Missouri River from Fort Atkinson to sign treaties with the Arikara, Cheyenne, Crow, Mandan, Ponca and several Sioux tribes. Atkinson was also commander of the military arm of the Yellowstone Expedition of 1818-19. He died at Jefferson Barracks near St. Louis and was buried there. Fort Atkinson, Wisconsin, and Fort Atkinson, Iowa, are both named for him. Heitman, *HDRUSA*, vol. 1, p. 174; Julius Stirling Morton, *Illustrated History of Nebraska*. Lincoln: Jacob North & Company, 1907. Vol. II, p. 140, n.1.

(218) French Mills, Franklin County, New York, has been called Fort Covington since 1817. Leonard Covington (1768-1813) of Maryland joined the regular army in 1792 as a Cornet of Cavalry, was promoted to Captain, then resigned in 1795 to serve in the Maryland State Legislature and U.S. Congress. He re-entered the army and was promoted to Brigadier General in 1813. Covington led a brigade in Wilkinson's army and was shot in the stomach at Crysler's Field, dying 3 days later at French Mills. Known as an energetic and competent leader, his loss, wrote one historian, "was a tragedy for the army." Graves, 4. (reference to the article, "Hard School ...", citation #99)

(219) Albon Man, born in Connecticut in 1769, settled at French Mills, then Constable, New York, where in addition to practicing medicine he operated a farm and sawmill, and acted as Town Supervisor. He was a founder of both the Franklin and Clinton County Medical Societies (1807, 1809) and was put on General Wilkinson's staff after the Battle of Crysler's Field due to the shortage of physicians. Man was a dedicated medical practitioner, sometimes travelling 40-50miles to treat patients (in one instance he went as far as Montreal), and died in 1820 after being thrown from his horse after treating a seriously ill patient at Fort Covington. His home at Constable, the "Man Homestead", is preserved as a New York State Historic Landmark. Neil B. Yetwin, "Dr.Albon Man: 'A Physician of Large Practice'." *Franklin County Historical Review* (Malone, New York: The Franklin County Historical Society & Museum), Vol. 26, 1989. pp. 22-32.

(220) Buel H. Hitchcock (d. December 4, 1836) came to Fort Covington in 1802 from Shoreham, Vermont, having travelled part of the way with Albon and Alric (Albon's brother) Man. Hitchcock, a physician and Freemason, owned and operated the "Hitchcock House" at the south foot of Water Street. Seaver, 317; Duane Hamilton Hurd, *History of Clinton and*

Franklin Counties, New York (Philadelphia: J.W.Lewis & Co., 1880), p. 481. Hereafter cited as Hurd.

(221) Charlotte Bailey was born on October 12, 1796, at Plattsburgh, New York, to Judge William Bailey and his first wife, Hannah Hagerman. She spent the years 1800-1811 at her father's homestead in Chateaugay and was sent to a girls' school in Montreal until war was declared in 1812. A refined, affectionate and self-reliant young woman who was described as slender, with blue eyes and brown hair, she was sent to her uncle Albon Man's homestead at Constable to avoid any contact with soldiers stationed there who were preparing for war. *BSBMMF,* pp. 17-21.

(222) A town in Franklin County, New York, formed from the town of Chateaugay in 1805. A hospital with 450 sick and wounded was set up in Malone in squalid conditions after the Battle of Crysler's Field, while Wilkinson stayed there to convalesce in a private home. "Malone", in Seaver, Chapter XXV, et passim.

(223) Myers remained at Man's home from November 15, 1813, to March, 1814. He then followed the army to Plattsburgh, to where it had retreated after abandoning French Mills. Albon Man, *No. 3 Accts Current, Constable, NY.* Courtesy of the family of Richard W. Floyd.

(224) Mordecai Myers and Charlotte Bailey were married at the Man Homestead on January 14, 1814. *Plattsburgh Republican,* January 29, 1814, p. 3, Col. 4. William Bailey was born in Poughkeepsie, New York, on November 11, 1763, and served in the Dutchess County Militia during the Revolution. He operated a grocery business in New York City for a time, before moving to Plattsburgh in 1786, where he owned another store and became First Major in a militia regiment commanded by his brother-in-law, Lieutenant Colonel Benjamin Mooers. He moved his family for several years to a tract in Chateaugay, where he operated a farm, forge and paper mill. After moving back to Plattsburgh, he served as one of the Associate Justices of the Clinton County Court of Common Pleas, represented Clinton and Essex Counties in the New York State Assembly, became the first Judge of the County, and was one of Plattsburgh's original 8 Trustees. During the War of 1812, Bailey and partner Henry DeLord operated the "Red Store," the only one in Plattsburgh to extend credit to officers and men stationed there. General Macomb had promised that the army's credit certificates were "as good as gold," but once the army withdrew and the partners' goods exhausted, Congress refused to reimburse the men, and the business went bankrupt. Bailey tried unsuccessfully for the next 25 years to recoup their losses. He died in 1841. *BSBMMF,* pp. 17-18; Allan S. Everest, *The War of 1812 in the Champlain Valley* (Syracuse, New York: Syracuse University Press, 1981), p. 156, pp. 202-203.

(225) Myers incorrectly refers here to the court-martial of General Edmund Pendleton Gaines (1777-1849), who was court-martialed at the start of the Mexican War by calling up Louisiana Volunteers. *DAB*, IV, pp. 92-93; Heidler, pp. 199-200, Heitman, *HRUSA*, p. 281. Myers meant the court-martial of Colonel Isaac Coles of the 12th Infantry Regiment. Coles had been challenged to a duel by Dr. James C. Bronaugh, a hospital surgeon, whom Coles court-martialed for living with the wife of a common soldier. Bronaugh publicly declared Coles "a base liar, infamous scoundrel, and coward." Coles refused to participate in the duel and had Bronaugh arrested. A court-martial subsequently convicted Bronaugh but he was retained in the army and became one of Andrew Jackson's closest associates in Florida (see citation #144, and Alan Taylor, *The Civil War of 1812, British Subjects, Irish Rebels & Indian Allies* (New York: Alfred A. Knopf, 2010), p. 324, n. 17, n.18; Myers was at Sackett's Harbor when he received a November 16, 1814, order from General Jacob Brown's Adjutant General C.K. Gardner at Buffalo to attend a General Court Martial in New York City as a witness "on the trial of Col I A Coles." (Adj. Gen. C.K. Gardner to Myers, November 16, 1814. MMC, Clements Library). Coles' trial convened at Governor's Island, New York Harbor, on December 15, 1814. Hastings, *Tompkins Papers*, vol. 1, p. 728.

(226) Odelltown is now part of LaColle, a former town in Southern Quebec named after United Empire Loyalist Joseph Odell of Poughkeepsie, New York. In November 1812, 300 regulars and 200 militia under General Dearborn nearly reached Odelltown but were stopped by General De Salaberry's 2 companies of Canadian Voltigeurs, 300 Native allies and some local militia. General Wade Hampton's 5000 troops entered Odelltown in Spetmber 1813, stayed for 2 days, then marched toward Chateaugay with the intention to attack Montreal, but they, too, were driven back by De Salaberry. In March 1814 General Wilkinson's 5000 troops planned to attack Odelltown and LaColle, but were turned back by Canadian forces. John A. Bilow, *Chateaugay, N.Y and The War of 1812* (St-Lambert,Quebec: Payette & Simms, 1984), p. 114.

(227) Named for Captain de Chazy, who was killed by 60 Mohawks in 1664, Chazy is a township of 53 square miles bordered on the east by Lake Champlain. The town's chief importance during the War of 1812 was that it lay in the path of thousands of soldiers from invading armies, destroying businesses, property and growth. During the summer of 1814, many Chazy and Champlain residents moved to Peru, New York, for safety because of the threats of impending battles. Nell Jane Barnett and David Kendall Partin, *A History of the Town of Chazy – Clinton County, New York* (Burlington, Vermont: George Little Press, Inc. 1970), p. 12, pp. 97-102.

(228) Jacob Brown (1775-1828), born to a Pennsylvania Quaker family, was a teacher, surveyor, lawyer and secretary to Alexander Hamilton who founded Brownsville, New York, near Sackett's Harbor. He became a county judge, Colonel of Militia and Brigadier General, and when war was declared in 1812 was appointed commander of the 200-mile frontier from Oswego to Lake Francis. Brown was wounded twice during the war, received thanks and a gold medal from Congress, and in 1821 was made General-in-Chief of the Northern Division of the U.S. Army. He died in Washington in 1828. *DAB*, II, 124-126; Heitman, *HRUSA*, p. 150; see *John D. Morris, Sword of the Border: Major General Jacob Jennings Brown, 1775-1828* (Ashland, Ohio: Kent State University Press, 2000), et passim.

(229) The siege of Fort Erie, which began on July 3, 1814, was lifted on September 21st. The Battle of Chippawa occurred on July 5, 1814, and the Battle of Lundy's Lane on July 25, 1814.

(230) Alexander Macomb (1782-1841) served in the militia and was one of the first graduates of West Point. He became Lieutenant Colonel of Engineers and Adjutant-General of the Army and promoted to Brigadier General effective January 1, 1814. When General Izard withdrew from the military post on Lake Champlain in the summer of 1814, Macomb was left in command of that region. He won a victory over the British at the Battle of Plattsburgh (September 11, 1814) and was commissioned a Major General, in addition to receiving thanks and a gold medal from Congress. Macomb was appointed General-in-Chief of the Northern Division of the Army upon the death of General Jacob Brown. *DAB*, VI, 155-157.

(231) Myers made several trips to St. Regis on spy missions dressed in civilian clothing. Graves, *FOG*, p. 328.

(232) Myers was discharged from the Army on June 15, 1815 (official date of record). Heitman, *HRUSA*, p. 740. However, he was retained in active service and credited with pay until September 10, 1815.

(233) Myers meant "General," not "Colonel" in reference to Macomb.

(234) Robert Brent (1764-1819) of Maryland and Washington, D.C., was Paymaster of the Army (Paymaster-General) from July 1, 1808, until his resignation on August 28, 1819; he died several weeks later on September 7. Heitman, HDRUSA, vol. 1, p. 242; HRUSA, p. 143; Moser, et al, V, p. 20. For a correspondence regarding Myers's settling of accounts, see letter: "Mr. Tobias Lear of the Dept. of War to Mr. R. Brent, relating to Cap M. Myers of 13th Inf, 1815." PID 365144; Call # P464, courtesy AJHS. Tobias Lear (1762-1816) was personal secretary to President George Washington, and served as Thomas Jefferson's envoy to Saint-Domingue, and peace envoy in the Mediterranean during the First and Second Barbary Wars.

(235) John Palmer (1785-1840) was a lawyer, militia officer and original trustee of Plattsburgh,New York. Palmer went on to serve as District Attorney of Clinton County, New York, a New York State Assemblyman, Judge of the Clinton County Court of Appeals, and United States congressman from the 19th Federal District of New York, which included Clinton, Essex, Franklin and Warren Counties. He died on the island of St. Bartholomew in the French West Indies. Hurd, pp. 124-125; Dodge, p. 1696.

(236) It is likely that this friend was Naphtali Phillips (1773-1870), who was prominent in Democratic circles and political committees in New York City. After 1801 he was the proprietor of the Democratic-Republican journal, the *National Advocate,* and attaché of the New York Custom House. Phillips was the uncle of Mordecai M. Noah, once editor of the *National Advocate* and best known for his unsuccessful attempt to establish a refuge for Jews, "Ararat," on Grand Island in the Niagara River. At age 98, Phillips was the oldest surviving member of the Tammany Society and served as president of New York Congregation Shearith Israel (where Myers had served in several positions prior to the war) for 14 terms. Myers occasionally partnered with Phillips in his brokerage and auctioneering concerns before and after the war. "Naphtali Phillips", *PAJHS,* Vol. XXI (1913), pp. 172-232.

(237) Charles Mitchell had been a 3rd Lieutenant in the *13th Infantry* and was honorably discharged on June 15, 1815. He came to New York City in August 1815 seeking Myers's help. Myers lent him money and hired him as a clerk in his auctioneering and brokerage business. Mitchell stole into Myers's office on October 17, 1815, broke into his desk, robbed $50 in gold coins and 17 Medical Science Lottery tickets, and sent Myers threatening letters. He also forged checks using the name of Dr. Daniel Lord. Mitchell was arrested and confined on January 26, 1816, and ordered to leave the state. Heitman, *HRUSA,* 470; *The New York City-Hall Recorder, For March, 1816* (New York: Clayton & Kingsland, 1816), Vol. I, No. 3, pp. 41-43.

(238) Myers meant Dr. Daniel Lord, Jr. (1795-1845), a physician and druggist of New York City. Marynita Anderson Nolosco, "Physician Heal Thyself: Medical Practitioners of eighteenth Century New York." *American University Studies, Series IX, History: Vol. 170* (New York: Peter Lang Publishing, Inc. 2004), p. 169; *Memorial of Daniel Lord* (New York: D. Appleton and Company, 1869), pp. 45-5.

(239) Jacob Radcliff (1764-1841) was a graduate of Princeton and a founder of Jersey City, New Jersey. He practiced law mainly in Poughkeepsie, New York, but also in New York City. Radcliff served on the New York State Supreme Court, was the Mayor of New York City, and was a delegate to the Constitutional Convention at Albany in 1821. He and James Kent revised

many of New York State's laws. James Terry White, ed. *National Cyclopedia of American Biography* (Clifton, NJ: J.T. White & Co., 1891), Vol. 13, p. 471.

(240) An Ohio native, attorney Joseph Watson (d. October 7, 1836) was a District Paymaster of the army beginning July 25, 1812, until his honorable discharge on June 15, 1815. Watson, with Myers, established an agency that helped soldiers who did not want or could not get land warrants and patents for military bounty lands. The partners would "procure the land warrant, disgnate and draw the land by lot, and obtain and transmit the patent." Heitman, *HDRUSA,* vol. 1, 1009; *National Intelligencer,* Washington, D.C., September 19, November 7, 1816, quoted in Alina Marie Lindegren, *A History of the Land Bonus of the War of 1812,* (University of Wisconsin. Unpublished Masters Thesis, 1922), p. 17.

(241) James Vanderpoel practiced law in Kinderhook and Kingston, New York, served as a New York State Assemblyman, artillery Captain in the War of 1812, County Surrogate, Judge of the County Court, and Circuit Judge of the Supreme Court of New York State. He sold the Kinderhook property to Myers when he moved to Albany in 1843. James Edward Leath, *The Story of the Old House: Columbia County's "House of History"* (Kinderhook, N.Y.: The Columbia county Historical Society, 1947), pp. 11-12.

(242) See the editor's introduction for Myers's six one-year terms in the New York State Assembly.

(243) Myers recalled the date incorrectly. He was elected "President" (mayor) of Kinderhook on May 9, 1839. *Kinderhook Sentinel,* May 9, 1839, p. 2.

(244) See the editor's introduction for Myers's participation in the festivities celebrating Van Buren's visit and those of the outgoing President's return to Kinderhook marking his retirement from public affairs.

(245) This was Lieutenant Thomas Vail. (see note #57)

(246) This location was probably Smith M. Salisbury's Printing Office, 2 blocks from Pomeroy's Tavern (where officers sometimes boarded) and McEwan, Shoemaker, on Seneca Street, where Myers likely boarded. (See notes #67 and #138)

(247) This is a reference to the "Adams" and "Caledonia" (See note #83)

(248) Jesse Duncan Elliott. (See note #84)

(249) Josiah Tatnall (1796-1871) was a grandson of Georgia Governor Tatnall and graduate of Norwich University at Atlanta. He entered the Navy as a Midshipman and participated in engagements with the British fleet near Croney's Island, Virginia (June 22, 1813), and was a volunteer at the Battle of Bladensburg, Maryland (August 24, 1814). After the war he served on the *U.S.S. Macedonia* off the coast of South America, where he

wounded a British officer in the shoulder after a duel over the issue of the American naval victories of 1812. He was a commander with Decatur during the Algerian War, participated in the bombardment of Vera Cruz in the Mexican War, and in 1860 transported Japanese ministers to the United States. Tatnall joined the Confederacy when the Civil War commenced and was in command of vessels at Norfolk, VA when the ironclad "Merrimac" was destroyed. He ended his career commanding the so-called "mosquito fleet" at Savannah, Georgia. William Arba Ellis, ed., *Norwich University 1819-1911. Her History, Her Graduates, Her Roll of Honor* (Montpelier, Vt: The capital City Press, 1911), p. 402; Lossing, p. 615, p. 680; *United States Naval Institute Proceedings.* Volume 35 –Part 2 (Annapolis – Maryland: U.S. Naval Institute, 1909), p. 1175.

(250) (See note #70) The march from Greenbush followed the route to Albany, Schenectady and Utica, and on to Buffalo and the Niagara Frontier.

(251) This incident explains the origin of the term "jolly snorters" when describing the *13th Infantry.* (See note #211)

(252) George Mercer Brooke, a native of Virginia, entered the army as a First Lieutenant of the *5th Infantry* on May 3, 1808, then Captain on May 1, 1810. He was promoted to Major of the *23rd Infantry,* May 1, 1814; transferred to the *4th Infantry,* May 17, 1815; transferred to the *8th Infantry,* January 27, 1819; promoted to Lieutenant Colonel, March 1, 1819; transferred back to the *4th Infantry,* August 13, 1819, and Colonel of the *5th Infantry,* July 15, 1831. Brooke was breveted Lieutenant Colonel on August 15, 1814, for gallant conduct in the defense of Ft. Erie, U[pper] C[anada], Colonel on September 17, 1814, for distinguished & meritorious service in the sortie from Ft. Erie; Brigadier General, September 17, 1824, for 10 years faithful service in one grade, and Major General on May 30, 1848, "for meritorious Service particularly in the performance of his duties in prosecuting the war with Mexico." Brooke died March 9, 1851. Heitman, *HDRUSA,* Vol. 1,248; *HRUSA,* 147.

(253) Virginia native Thomas Parker (1753-1820) was an officer in the Revolution and a Lieutenant Colonel in the regular army during the Quasi-War with France in 1799. He commanded the *12th Infantry* as a Colonel on the Niagara Frontier 1812-13, was promoted to Brigadier General in March 1813, and resigned from the army in March 1814. See Graves, "The Hard School of War....", p. 50.

(254) This officer is not able to be identified.

(255) Unable to identify this soldier with the information provided.

(256) The village of Champlain, New York, in Clinton County, is located in the Town of Champlain, north of Plattsburgh. It became part of the United

States in 1783 and was an important staging point during the War of 1812. This is not to be confused with The District of Northfield, formed in Ontario County, New York, in 1792. It became the town of Northfield in 1796, then was subdivided into Pittsford and Henrietta in 1844. It is 8 miles from Rochester, New York, in the southeastern portion of Monroe County, and is nowhere near Lake Champlain.

(257) Unable to identify this solder with the information provided.

(258) Homer Virgil Milton. (See note # 155)

(259) Thomas Beverly Randolph. (See note # 147)

(260) James Crane Bronaugh. (See note #144)

(261) John Stannard (See note #145)

(262) Pennsylvania native Richard Whartenby was a First Lieutenant in the *5th Infantry* on May 3, 1808, Captain on May 3, 1810, and Major of the *40th Infantry* on May 1, 1814. He was retained May 17, 1815, as Captain (permanent rank) of the *7th Infantry* with a Brevet to Major retroactive to May 1, 1814. Whartenby was promoted to Major (permanent rank) of the *1st Infantry* on April 30, 1817, and died on May 14, 1825. Heitman, HDRUSA, vol. 1, p. 1022; HRUSA, p. 686. Myers was a "second" or trusted friend, confidant, arbiter, mediator or "custodian of honor" who acted on behalf of the principal. See Allison L. La Croix, "To Gain the Whole World and Lose His Own Soul: Nineteenth-Century American Dueling As Public Law and Private Code". *Hofstra Law Review*, Vol. 33, 501 (204), pp. 517-518.

(263) Myers served 2 one-year terms in 1851 and 1854. (See the editor's introduction covering his activities as mayor.)

(264) Schenectady *Reflector & Democrat*, October 19, 1860, p.3.

(265) Proceedings of the Grand chapter Royal Arch Masons of New York State at Its One Hundred and Seventeenth Annual Convention Held Albany (New York: The Grand Chapter, 1914), p. 111; Robert W. Reid, *Washington Lodge No. 21, F. & A.M. and Some of its Members* (Washington Lodge Publishers, 1911), p. 163.

(266) Schenectady, New York had no newspaper by this name. It was the *Schenectady Evening Star* of January 25, 1871, p.3, that printed an extensive account of Myers's life; another local paper, the *Schenectady Daily Union*, published an even longer account of Myers's life and his funeral on January 24-25, 1871. Both issues on p. 3 respectively. Edited versions of the Schenectady pieces appeared in the Albany *Argus* of January 20, 1871, and *The New York Times* of January 25, 1871.

Theodorous Bailey Myers, Mordecai Myers's eldest surviving son to whom the letters comprising the Reminiscences were addressed.
Catalina Mason Myers Julian-James, *Biographical Sketches of the Bailey-Myers-Mason Families, 1776-1905* (Washington, D.C.: 1908). Hereafter cited as *Biographical Sketches. Photo courtesy of the SCHS.*

Mordecai Myers's commission as Captain in the
13th United States Infantry, dated July 6, 1812.
When viewed in its full size (14 inches wide by 17 inches
tall), Myers's name can be read on the fourth line
from the top. *Courtesy of the late Dr. John S. Hoes.*

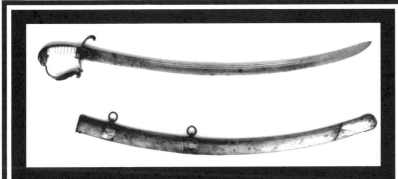

Mordecai Myers's sabre, used by him throughout the War of 1812, and donated to the National Museum in 1923 by his granddaughter, Mrs. Julian James. For a full description of the French-made sabre, see Theodore T. Belote, "American and European Swords"; *The Historical Collections of the United States National Museum* (Washington, D.C.: United States Government Printing Office, 1932), Bulletin #9, p. 33.

New York State historic marker in front of the Man Homestead, where Myers was taken for treatment of wounds he sustained at the Battle of Crysler's Field, November 11, 1813. *Photo, courtesy of Joyce Denny.*

The Man Homestead, Constable, New York. The window shown at the lower right on the first floor was where Myers was treated by Dr. Albon Man, and where Charlotte Bailey covered the window to keep sunlight from Myers's eyes while the doctor removed 30 fragments of bone from the wounded officer's shoulder. *Photo, courtesy of Joyce Denny.*

Miniature of Charlotte Bailey Myers at age 17, when she met Mordecai Myers and helped with his recovery. The original miniature has not been located, but a photograph of it appears in the *Biographical Sketches. Photo, courtesy of the SCHS.*

Above Left: Charlotte Bailey Myers at about age 37 or 38. *Courtesy, William Clements Library, University of Michigan, Ann Arbor, Michigan; Mordecai Myers Collection.*

Above Right: Charlotte Bailey Myers at about age 48, four years prior to her death from Consumption (Tuberculosis of the lungs) in New York City. *Image, courtesy of the late Dr. John S. Hoes.*

Judge William Bailey of Plattsburgh, New York; Mordecai Myers's father-in-law. *Biographical Sketches. Photo, courtesy of the SCHS.*

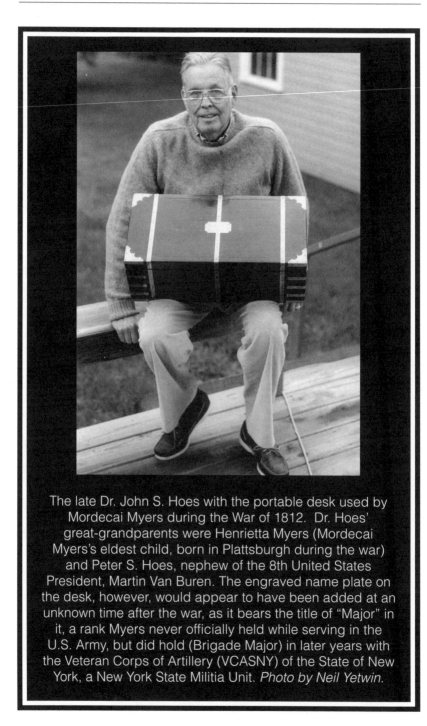

The late Dr. John S. Hoes with the portable desk used by Mordecai Myers during the War of 1812. Dr. Hoes' great-grandparents were Henrietta Myers (Mordecai Myers's eldest child, born in Plattsburgh during the war) and Peter S. Hoes, nephew of the 8th United States President, Martin Van Buren. The engraved name plate on the desk, however, would appear to have been added at an unknown time after the war, as it bears the title of "Major" in it, a rank Myers never officially held while serving in the U.S. Army, but did hold (Brigade Major) in later years with the Veteran Corps of Artillery (VCASNY) of the State of New York, a New York State Militia Unit. *Photo by Neil Yetwin.*

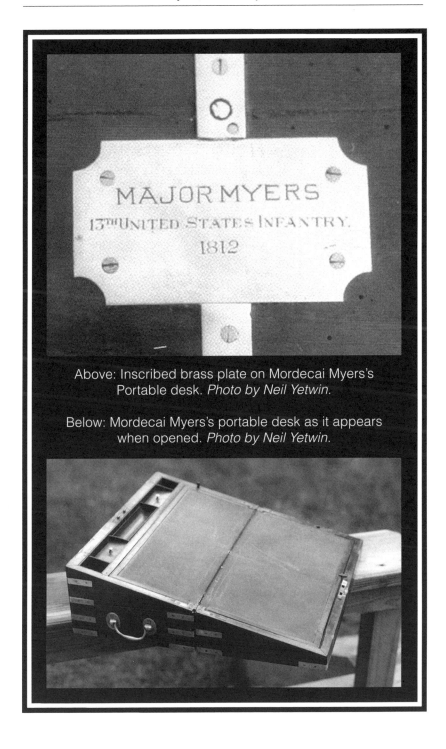

Above: Inscribed brass plate on Mordecai Myers's Portable desk. *Photo by Neil Yetwin.*

Below: Mordecai Myers's portable desk as it appears when opened. *Photo by Neil Yetwin.*

Above: The Myers Family homestead in Kinderhook, New York. The Myers lived there from 1834-35 to 1843 when they returned to New York City. It is now the James Vanderpoel House, and is maintained as part of the Columbia County Historical Society. *Courtesy, Columbia County Historical Society.*

Below: Mordecai Myers lived in the residence to the left, at 229 Union Street, Schenectady, New York, from 1865 to his death in 1871. His daughter, Francis, and son-in-law, Edgar Jenkins, lived next door at 231 Union Street, to the right of #229. *Photo by Neil Yetwin.*

Pencil sketch of Mordecai Myers, by New York City engraver William Barritt, brother-in-law of historian Benson J. Lossing. Lossing obtained an account of Myers's experiences during the war, and may have asked Barritt to visit Schenectady. The sketch appears to contain elements of the Jarvis portrait and an earlier miniature of Myers at age 21. In any case, the sketch was not used in Lossing's *Pictorial Field-Book of the War of 1812*. The original is part of the Bisno Papers collection of the American Jewish Historical Society (AJHS), and was located by Susan Woodland and Heather Halliday of the AJHS, who alerted the editor of its existence. *Courtesy, American Jewish Historical Society.*

A miniature of Mordecai Myers by Elkanah Tisdale, circa 1797-98. It was a gift from Myers to his mother, Rachel. The original has not been located, but a photograph appears in *Biographical Sketches*. *Image, courtesy of the SCHS*.

Portrait of Mordecai Myers by John Wesley Jarvis, 1810.
The two men served together in the New York State Militia
prior to the War of 1812.
Courtesy, Toledo Museum of Art; 1950.275.

Mordecai Myers, in a copy of the Jarvis portrait done by
Schenectady artist Samuel Sexton, circa 1860.
Courtesy of the late Dr. John S. Hoes.

Mordecai Myers, in a copy of the Jarvis portrait (unframed); artist and date unknown. Currently on display in the dining room of the James Vanderpoel House, Kinderhook, New York, this painting was gifted to the Columbia County Historical Society in 1928 by Miss Fannie Jackson, Mordecai Myers's granddaughter. *Courtesy, the Columbia County Historical Society (CCHS).*

Charcoal sketch of Mordecai Myers; artist unknown.
Courtesy of the American Jewish Historical Society.

An unknown Lieutenant in the *Veteran Corps of Artillery of the State of New York*, c. 1818.

The *Veteran Corps of Artillery of the State of New York (VCA)* was organized on November 25, 1790, in New York City and approved as a separate and distinct organization in the Militia of the State of New York by Governor George Clinton on March 6, 1791, and as a separate Corps in the Active Militia by Act of Congress on May 8, 1792. The *VCA* was the first New York State unit to volunteer for federal duty in the War of 1812, and was called into serve in 1812 and 1814. Mordecai Myers joined the *VCA* after the war, and served as Brigade Major from 1825-1835.

Frontispiece Image from book: *Roster of the Veteran Corps of Artillery, Constituting The Military Society of the War of 1812 for 1901-1902* (New York City: The Veteran Corps of Artillery, Adjutant's Office, 1901).

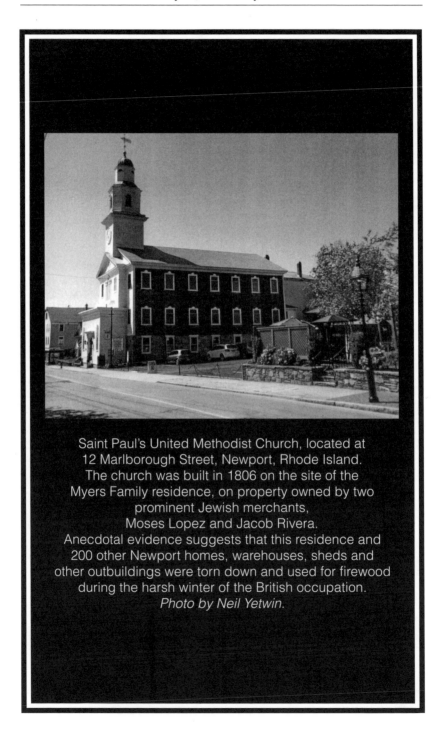

Saint Paul's United Methodist Church, located at
12 Marlborough Street, Newport, Rhode Island.
The church was built in 1806 on the site of the
Myers Family residence, on property owned by two
prominent Jewish merchants,
Moses Lopez and Jacob Rivera.
Anecdotal evidence suggests that this residence and
200 other Newport homes, warehouses, sheds and
other outbuildings were torn down and used for firewood
during the harsh winter of the British occupation.
Photo by Neil Yetwin.

The Myers Family burial plot marker, Vale Cemetery, Schenectady, New York. *Photo by Neil Yetwin.*

"M. M.
Severley wounded at Chrysler's Field in Canada, when Capt. of the 13th U.S. Infantry, in the war of 1812. Represented New York City for many years in the State Legislature. Mayor of this City [Schenectady] & Grand Master of Masons of the State of New York.
The Historic Pen hath writ his spotless name Among the good and brave, the men of deathless fame.

Appendix:
Mordecai Myers and the Founding of
The Military Society of the War of 1812

The Military Society of the War of 1812 (organized as the "Society of the War of Eighteen Hundred and Twelve") was instituted January 3, 1826, by commissioned officers and ex-officers of the armies and navies (regulars and volunteers) of the United States in the War of 1812 who formed an Association to petition Congress for bounty and land rights, as was accorded to veterans of the Revolution, by the Federal and States governments. In September 1826, their petition having been presented to Congress, the association determined to remain together as a Military Society; permanent officers were elected and organizational rules were adopted.

Mordecai Myers was chosen at the original January 3, 1826, meeting to serve as Secretary, and his role as an organizing founder was documented in an 1895 publication of the Society: *The Military Society of the War of 1812: Annals, Regulations, and Roster* (Secretary and Adjutant's Office, March 12, 1895. Copyright, 1895, by The Society of the War of 1812). His military rank and unit was described as: "Captain Mordecai Myers, formerly 13th Regiment, U. S. Infantry..." and he is referred to repeatedly as "Captain Myers."

"ANNALS.

"On January 3, 1826, pursuant to notice published in the newspapers of the City of New York of December 31, 1825, a number of commissioned officers and ex-officers, who resided or were stationed in the vicinity, and who had served with reputation in the Army of the United States in the War of 1812, met at the Broadway House, corner of Broadway and Grand Streets, 'to take into consideration the expediency of presenting a respectful petition to Congress, praying for a grant of public lands, agreeable to rank and former practice, as a reward for their services, sufferings, and losses during the Second War of Independence.'

"Major George Howard, formerly 1st Regiment, U. S. Infantry, was called to the Chair; Captain Mordecai Myers, formerly 13th Regiment, U. S. Infantry, was chosen Secretary, and Major Clarkson Crolius, formerly 27th Regiment, U. S. Infantry, was appointed Treasurer. The meeting then, after due consideration, unanimously resolved that it was expedient to present to Congress a respectful memorial, praying for lands as a reward for past services.

"A committee was, thereupon, appointed to draft and forward the memorial as called for by the action of the meeting. This committee consisted of Brigadier General, the Honorable Robert Bogardus, formerly Colonel 41st Regiment, U. S. Infantry; Captain Mangle Minthorne Quackenboss, late 8th Regiment, U. S. Infantry, together with Majors Howard and Crolius and Captain Myers.

"The committee met on January 8, 1826, the anniversary of the battle of New Orleans, at the residence of Captain Myers, No. 45 Mercer Street, in the City of New York, and prepared the memorial to which Colonel Joseph Watson, formerly District Paymaster U. S. A., Colonel Joseph Lee Smith, late 3d Regiment, U. S. Infantry, Colonel Gilbert Christian Russell, formerly 20th Regiment, U. S. Infantry, and Colonel James R. Mullany, formerly 32d Regiment, U. S. Infantry, and then Quartermaster General U. S. A., who had been added to the committee after the meeting, also affixed their signatures.

Bibliography

Newspapers:

Newspapers (Introduction and Family History)

Albany Argus (Albany, New York, January 20, 1871).

Albany Evening Journal (Albany, New York, January 3, 1832).

Connecticut Courant, September 17, 1787: Issue 1182.

[Kinderhook] Sentinel (Kinderhook, New York)
October 26, 1837	May 9, 1839	July 18, 1839
July 23, 1840	November 7, 1845	February 15, 1848

Litchfield Monitor/The Weekly Monitor, September 24, 1787, Issue 144.

Middlesex Gazette, September 24, 1787, Vol. II, Issue 99.

National Intelligencer (Washington, D.C.)
September 19, 1816   November 7, 1816).

National Intelligencer (Washington, D.C., November 7, 1816).

New York Enquirer, "M. Myers & Co., 107 Water St.", (New York, New York, July 7, 1826).

New York Morning Courier, (New York, New York, November 6, 1828).

New York Times (New York, New York)
January 25, 1871   August 14, 1912; SM14

Plattsburgh Republican (Plattsburgh, New York)
July 3, 1812   July 24, 1812   January 29, 1814

Schenectady Daily Union. (Schenectady, New York)
January 24, 1871;   January 25, 1871

Schenectady Evening Star (Schenectady, New York)
January 7, 1855	January 20, 1871	January 21, 1871
January 23, 1871	January 25, 1871	

Schenectady Reflector (Schenectady, New York)
May 19, 1848	September 6, 1850	April 14, 1851
April 11, 1851, to April 8, 1852, time spread		March 31, 1854
April 7, 1854	December 18, 1857	

Schenectady *Reflector & Democrat,* (Schenectady, New York)
October 19, 1860	July 16, 1863	October 8, 1863
February 7, 1865).		

Schenectady Union Star (Schenectady, New York, December 31, 1914).

Other Published Sources:

Adler, Selig. *From Ararat to Suburbia: The History of the Jewish Community of Buffalo.* (Philadelphia: The Jewish Publication Society of America, 1960).

Allen, William. *The American Biographical Dictionary* (Boston: John P. Jewett and Company, 3rd ed., 1857).

Allison, Charles Elmer. *The History of Yonkers* (New York: W.B. Ketcham, 1896).

American Ancestry: Giving the Name and Descent, in the Male Line of Americans Whose Ancestors Settled in the United States Previous to the Declaration of Independence, A.D. 1776 (Albany, N.Y.: Joel Munsell's Sons, 1894), Volume 9.

American Jewish Historical Society. "Items Relating to the Jews of Newport" and "Items-Seixas Family". *Publications of the American Jewish Historical Society,* Vol. XXVII (1920).

American Jewish Historical Society. "Registry of circumcisions by Abm. I. Abrahams From June 1756 to January, 1783 in New York, in Hebrew and English" in "Misc. Items, N.Y. Congregations". *Publications of the American Jewish Historical Society,* Vol. XXVII (1920).

American Masonic Record and Albany Saturday magazine, Vol. I, No. 10, April 7, 1827; p.78.

Annual Reports of the War Department For the Fiscal Year June 30, 1900. Reports of Chiefs of Bureaus (Washington: Government Printing Office, 1900).

Appleton's Cyclopedia of Biography (New York: D. Appleton and Company, 1888).

Babcock, Louis L. *The War of 1812 on the Niagara Frontier* (Buffalo, New York: Buffalo Historical Society, 1927).

Baum, Charlotte; Hyman, Paula; & Michel, Sonya. *The Jewish Woman in America* (New York: The Dial Press, 1975).

Baynard, Samuel Harrison. *History of the Supreme Council, 33: Ancient Accepted Scottish Rite of Freemasonry Northern Masonic Jurisdiction of the United States of America and Its Antecedents* (Boston, MA: Grit Publishing Company, 1938. Vol I).

Bailey, Theodorous. Theodorus Bailey to Judge William Bailey, April 3, 1823. *Mordecai Myers Collection,* Clements Library, University of Michigan, Ann Arbor, Michigan. Vol. II.

Bailey, William. Judge William Bailey to Mordecai Myers, November 8, 1821. *Mordecai Myers Collection,* Clements Library, University of Michigan, Ann Arbor, Michigan. Vol. II.

Bailey, W.T. *Richfield Springs and Vicinity. Historical, Biographical, and Descriptive* (New York and Chicago: A.S. Barnes & Company, 1874).

Barnett, Nell Jane; and Partin, David Kendall. *A History of the Town of Chazy – Clinton County, New York* (Burlington, Vermont: George Little Press, Inc. 1970).

Bassett, John Spencer; and Maydole, David Matteson, eds. *Correspondence of Andrew Jackson,* (Washington, D.C.: Carnegie Institution, 1926-1935), Vol. 2.

Bauer, K. Jack. *Zachary Taylor: Soldier, Planter, Statesman of the Old Southwest* (Baton Rouge, Louisiana: Louisiana State University Press, 1985).

Baxter, Maurice G. *Daniel Webster & the Supreme Court* (Amherst, MA: University of Massachusetts Press, 1966).

Bell, D.G. *Early Loyalist Saint John: The Origin of New Brunswick Politics 1783-1786* (Fredericton, New Brunswick: New Ireland Press, 1983).

Bell, William Gardner. *Commanding Generals and Chiefs of Staff: Portraits and Biographical Sketches* (Washington, D.C.: Government Printing Office, 1992).

Bilow, John A. *Chateaugay, N.Y. and The War of 1812* (St-Lambert, Quebec: Payette & Simms, 1984).

"Biography of Major-General Benjamin Mooers of Plattsburg, Clinton County, N.Y., Written in 1833, By Request of His Son, Benjamin H.Mooers." *The Historical Magazine and Notes and Queries Concerning the Antiquities of History and Biography of America,* Vol. I, third Series (Morrisania, N.Y.: Henry B. Dawson, 1872-73).

Blau, Joseph L., ed., *Cornerstones of Religious Freedom in America* (Boston: The Beacon Press, 1950).

Blau, Joseph L. and Baron, Salo W., eds., *The Jews of the United States 1790-1840: A Documentary History* (New York and London: Columbia University Press, 1963).

Boerstler, Charles. Charles Boerstler to [Mordecai] Myers, February 23, 1813. *Mordecai Myers Collection,* Clements Library, University of Michigan, Ann Arbor, Michigan.

Boerstler, Dr. Christian. "Autobiography of Bavarian Immigrant: 'I Was a Liberty-Loving Man'." *The Connecticut Magazine,* Vol. XII, Spring of 1908, Number 1.

Broches, S. *Jews in New England: Six Historical Monographs* (New York: Bloch Publishing Co., 1942).

Brock, Robert Alonzo and Lewis, Virgil Anson. *Virginia and Virginians: Eminent Virginians* (Richmond and Toledo: H.H. Hardesy, Publishers, 1888. Vol. I).

Brockerway, Charles A. *One Hundred Years of Aurora Grata, 1808-1908* (Brooklyn, N.Y.: Aurora Grata Consistory, 1908).

Brown, Imogene E. *American Aristides: A Biography of George Wythe* (Rutherford, NJ: Fairleigh Dickinson University Press, 1981), et passim.

Brown, Wallace and Senior, Hereward. *Victorious in Defeat: The Loyalists in Canada* (Toronto & New York: Methuen Publications, 1984).

Burrows, Edwin G., and Wallace, Mike. *Gotham: A History of New York City to 1898* (Cary, North Carolina: Oxford University Press, 1998).

Burstein, Andrew. *The Original Knickerbocker: The Life of Washington Irving* (New York: Basic Books, 2007).

Callahan, North. *Flight From the Republic: The Tories of the American Revolution* (New York: The Bobbs-Merrill Company, Inc., 1967).

Carnochan, Janet. "Fort Niagara," *Niagara Historical Society:* No. 23 (orig. pub. 1922).

Catalogue of Members Phi Beta Kappa. (New Haven, Connecticut: Connecticut Alpha [Yale University], 1905).

Census – City of Schenectady, 1850.

Census of Schenectady County, 1855.

Chadwick, Bruce. "The Mysterious Death of George Wythe," *American History,* February 2009.

Chambers, James A. Lt. James A. Chambers to Mordecai Myers, March 1, 1815. *Mordecai Myers Collection,* Clements Library, University of Michigan, Ann Arbor, Michigan.

Chyet, Stanley F. *Lopez of Newport: Colonial American Merchant Prince* (Detroit: Wayne State University Press, 1970), Et passim.

Clark, Hiram C. *A History of Chenango County* (Norwich, N.Y.: Thompson & Pratt, 1850).

Clark, Isaac. Isaac Clark to Mordecai Myers, September 2, 1812. *Mordecai Myers Collection,* Clements Library, University of Michigan, Ann Arbor, Michigan.

Cohen, Ira. "The Auction System in the Port of New York, 1817-1837." *The Business History Review.* Vol. 45, No. 4 (Winter, 1971).

Coles, Harry L. *The War of 1812* (Chicago and London: The University of Chicago Press, 1965).

Collier, Edward A. *A History of Old Kinderhook* (New York and London: G.P. Putnam's Sons, 1914).

Cruickshank, Ernest, ed. "Diary of John Keyes Paige, 13th Regiment" in *The Documentary History of the Campaign Upon the Niagara Frontier In the Year 1813. Part III (1813), August to October, 1813* (Welland, Ontario: The Lundy's Lane Historical Society).

Cruickshank, Ernest A., ed. *The Documentary History of the Campaign Upon the Niagara Frontier In The Year 1812, Part II* (Welland, Ontario: The Lundy's Lane Historical Society, n.d.).

Cullum, George W. *Biographical register of the officers and graduates of the U.S. Military Academy, at West Point, N.Y., from its establishment, March 16, 1802, to the army re-organization of 1866-67. By Bvt. Major-General George W. Cullum* (New York: D. Van Nostrand, 1868).

Cunliffe, Marcus. *Soldiers & Civilians: The Martial Spirit in America 1775-1865* (Boston & Toronto: Little, Brown and Company, 1968).

Curry, Daniel. *A Historical Sketch of the Rise and Progress of the Metropolitan City of America.* (New York: Carlton & Phillips, 1853).

Cushing, James S. *The Genealogy of the Cushing Family* (Montreal: The Perrault Printing Company, 1905).

Dandridge, Danske. *American Prisoners of the Revolution* (Charlottesville, VA: The Michie Company, 1911).

Davis, Matthew. *Memoirs of Aaron Burr, with Miscellaneous Selections From His Correspondence* (New York: Harper & Brothers, 1836), Vol. II.

Dean, John Ward; Folsom, George; Shea, John Gilmary. *The Historical Magazine and Notes and Queries Concerning The Antiquities, History and Biography of America*, Vol. II, third Series (Morrisania, N.Y.: Henry B. Dawson, 1873).

Deeds. Book 52. Office of the Schenectady County Clerk.

Description of Barracks at the Cantonment at Greenbush, Record Group No. 94, AGO Miscellaneous File, the National Archives, Washington, D.C.

Dexter, Franklin Bowditch, ed. *The Literary Diary of Ezra Stiles, D.D., LL.D.*, 3 volumes (New York: Charles Scribner's Sons, 1901), et passim.

Diary or Journal of Judge Thomas Palmer, Schenectady, NY, Jan. 1, 1853 to Dec. 31, 1853. Book IV. The Special Collections of the Schaffer Library, Union College, Schenectady, New York.

Dickson, Harold E. *John Wesley Jarvis: American Painter, 1780-1840. With a Checklist of His Works* (New York Historical Society, 1949).

Dictionary of American Biography (New York: Scribner, 1958-1964), Vol. I., Vol. II, Vol. IV, Vol. V, Vol. VI, Vol. VIII, Vol. IX, Vol. X, Vol. XVII.

Directory of Utica NY of Oneida County NY, 1817, n.p.

Documents of the Assembly of the State of New York, Fifty-Fifth Session, 1832 (Albany: E. Croswell, Printer to the State, 1832. Vol. III).

Dodge, Andrew R. *Biographical Dictionary of the United States Congress 1774-2005* (Washington, D.C.: U.S. Government Printing Office, 2006).

Dubeau, Sharon. *New Brunswick Loyalists: A Bicentennial Tribute* (Agincourt, Ontario: Generation Press, 1983).

Duffy, John; Hand, Samuel B.; Orth, Ralph H. *The Vermont Encyclopedia* (Lebanon, New Hampshire: University Press of New England, 2003).

Efner, Elijah D. "The Adventures and Enterprises of Elijah D. Efner: An Autobiographical Memoir." *Publications of the Buffalo Historical Society, Volume 4.* (Buffalo, New York: The Peter Paul Book Company, 1896).

Elliott, James E.; and Elliott, Nicko. *Strange Fatality: The Battle of Stoney Creek, 1813.* (Montreal, Quebec, Canada: Robin Brass Studio, 2009).

Ellis, Franklin. *History of Columbia County, New York* (Philadelphia: Everts & Ensign, 1878).

Ellis, William Arba, ed. *Norwich University 1819-1911. Her History, Her Graduates, Her Roll of Honor* (Montpelier, Vt: The capital City Press, 1911).

Elting, John R. *Amateurs, To Arms! A Military History of the War of 1812* (Chapel Hill, North Carolina: Algonquin Books, 1991).

Emerson, Edgar C., ed. *History of Cape Vincent,NY* (The Boston History Company, Publishers: 1898).

Encyclopedia Judaica "Ezra Stiles" (Jerusalem, Israel: Keter Publishing House Jerusalem Ltd., 1972. XV).

Encyclopedia Judaica. "Familiants Laws" (Jerusalem, Israel: Keter Publishing House Jerusalem Ltd., 1972. XI).

Encyclopedia Judaica. "Maria Theresa" (Jerusalem, Israel: Keter Publishing House Jerusalem Ltd., 1972. VIII).

Encyclopedia Judaica. "Meir" (Jerusalem, Israel: Keter Publishing House Jerusalem Ltd., 1972. XI).

Encyclopedia Judaica. "Richmond", (Jerusalem, Israel: Keter Publishing House Jerusalem Ltd., 1972. XIV).

Everest, Allan S. *The War of 1812 in the Champlain Valley* (Syracuse, New York: Syracuse University Press, 1981).

Ezekiel, Herbert T. and Lichtenstein, Gaston. *The History of the Jews of Richmond From 1769 to 1917* (Richmond, Virginia, 1917), et passim.

Ezekiel, Jacob. "The Jews of Richmond", *Studies in American Jewish History*, No. 4, 1894.

Fernow, Berthold. *Documents Relating To The Colonial History of the State of New York* (Albany: Wood Parsons & Co., Printers, 1887), Vol. 1.

Fowler, David J. "Benevolent Patriot: the Life and Times of Henry Rutgers, 1745-1830" (Special Collections and University Archives, Rutgers, The State University of New Jersey, 2010).

Fredericksen, John C. *The War of 1812 in Person: Fifteen Accounts by U.S. Regulars, Volunteers and Militiamen* (Jefferson, N.C.: McFarland & Company, Inc. Publishers, 2010).

Freeman, Bill. *Hamilton: A People's History* (Toronto, Ontario: James Lorimer & Company Ltd., 2001).

Gardner, C.K. Adjtant General C.K. Gardner to [Mordecai] Myers, November 16, 1814. *Mordecai Myers Collection*, Clements Library, University of Michigan, Ann Arbor, Michigan.

Graves, Donald E. *Field of Glory: The Battle of Crysler's Farm, 1813* (Toronto, ON: Robin Brass Studio, 1999).

Graves, Donald E. (ed.). *First Campaign of an A.D.C.:The War of 1812 Memoir of Lieutenant William Jenkins Worth, United States Army* (Youngstown, New York: Old Fort Niagara Association, 2012).

Graves, Donald E. "For want of this precaution so many Men lose their Arms: Official, Semi-Official and Unofficial American Artillery Texts, 1775-1815 Part 8: The Flood Tide of French Inflouence: The Work of Tousard and Duane, 1807-1810." *The War of 1812 Magazine*, Issue 15: May 2011.

Graves, Donald E. "Lawless banditti: Joseph Willcocks, his Canadian Volunteers and the Mutual Destruction on the Niagara during the Winter of 1813" [Part I], *Fortress Niagara* (Journal of the Old Fort Niagara Association): Vol. IX, No. 2, December 2007; and [Part II (Conclusion)] Vol. IX, No.3, March 2008.

Graves, Donald E. *Soldiers of 1814: American Enlisted Men's Memoirs of the Niagara Campaign* (Youngstown, New York: Old Fort Niagara Association, Inc., 1995).

Graves, Donald E. "The Hard School of War: A Collective Bibliography of the General Officers of the United States Army in the War of 1812." *The Napoleon Series – Military Subjects: The War of 1812 Magazine,* Issue 3. Part II: "the Class of 1813". June, 2006.

Graves, Donald E. *Where Right and Glory Lead! The Battle of Lundy's Lane, 1814* (Toronto, ON, Canada: Robin Brass Studio, Incorporated, 1997).

Greenbush Cantonment, Town of East Greenbush – Clinton Heights, N.Y., unpublished manuscript, n.d., courtesy of the East Greenbush Historical Society.

Greenleaf, Edward. *Genealogy of the Greenleaf Family* (Boston, 1896).

Guernsey, Rocellus S. *New York City and Vicinity During the War of 1812-15; Being a Military, Civic and Financial Local History of That Period* (New York: C.L. Woodward, 1889), vol. I.

Gutstein, Morris A. *The Story of the Jews of Newport: Two And A Half Centuries of Judaism, 1658-1908* (New York: Bloch Publishing Co., 1936).

Hamilton, Thomas. *Men and Manners in America* (Edinburgh: William Blackwood, 1833; Rpt. New York: Augustus M. Kelley, Publishers, 1968).

Hammond, John. *The History of Political Parties in the State of New-York, from the Ratification of the Federal Constitution to 1840* (Cooperstown, N.Y.: H & E Phinney, 1846), vol. I.

Hannay, James. *History of the War of 1812 Between Great Britain and The United States of America* (Toronto: Morang & Co. Limited, 1905).

Hastings, Hugh, ed. *Military Minutes of the Council of Appointment of the State of New York, 1783-1821* (Albany: James B. Lyon, 1901), Vol. II.

Hastings, Hugh, ed. *Public Papers of Daniel D. Tompkins, Governor of New York 1807-1817: Military* (New York and Albany: Wynkoop Hallenbeck Crawford, 1898), Vols. I & II.

Heidler, David S. and Jeanne T., eds. *Encyclopedia of the War of 1812* (Santa Barbara, California: ABC-CLIO, 1997).

Heitman, Francis B. *Historical Register and Dictionary of the United States Army From Its Organization, September 29, 1789 To March 2, 1903* (Washington: Government Printing Office, 1903. Vol. I).

Heitman, Francis B. *Historical Register of the United States Army, From Its Organization, September 29, 1789 to September 29, 1889* (Washington, D.C., 1890).

Heitman, Frances B. *Historical Register of Officers of the Continental Army During the War of the Revolution, April, 1778 to December, 1783* (Washington, D.C., 1893).

Hickey, Donald R. *The War of 1812: A Forgotten Conflict* (Urbana and Chicago: University of Illinois Press, 1989).

Higgenbotham, Dan. *Daniel Morgan: Revolutionary Rifleman* (Chapel Hill, North Carolina: University of North Carolina Press, 1961).

Historical Collections – Collections and Researches Made By the Michigan Pioneer and Historical Society (Lansing, Michigan: Wynkoop Hallenbeck Crawford Co. State Printers, 1905), Vol. XXXIV.

"History of the Durham Boat," *The Durham Township Historical Society Online*, 20 January 2012 (http://durhamhistoricalsociety.org/history2html).

History of the Great Lakes (Chicago: J.H. Beers & Co., 1899), Vol. I.

Hitsman, J. Mackay. *The Incredible War of 1812: A Military History*. Updated by Donald E. Graves (Toronto: Robin Brass Studio, 1999); originally published by University of Toronto Press, 1965.

Holt, Michael F. "The Antimasonic and Know Nothing Parties", in Schlesinger, Arthur M., ed., *History of U.S. Political Parties* (New York: Chelsea House Publishers, 1973. Vol. I.

Homans (Jr.), J. Smith, ed., *The Bankers' Magazine and Statistical Register*, January, 1862, vol. 16, Part 2.

Hough, Franklin Benjamin. *The New York Civil List*, (Albany, New York: Weed, Parsons and Co., 1860).

Howell, George Rogers; and Tenney, Jonathan. *History of the County of Albany, N.Y., From 1609 To 1886* (New York: W.W. Munsell & Co., Publishers, 1886).

Hudson–Mohawk Geneological and Family Memoirs (New York: Lewis Historical Publishing Company, 1911), Vol. II.

Huhner, Leon. "Jews in the War of 1812", *American Jewish Historical Society*, XXVII (1918).

Huhner, Leon. "The Jews of Virginia From The Earliest Times To the Close of the Eighteenth Century", *Publications of the American Jewish Historical Society*, Vol. XX (1911).

Hurd, Duane Hamilton. *History of Clinton and Franklin Counties, New York* (Philadelphia: J.W.Lewis & Co., 1880).

Hyatt, A. M. J. "Henry Procter." in *Dictionary of Canadian Biography Online*. Vol. 6, 1821-1835.(Toronto: University of Toronto Press, 1979).

Isenberg, Nancy. *Fallen Founder: The Life of Aaron Burr* (New York: Viking Penguin, 2007).

"Items Relating to the Jews of Newport," *Publications of the American Jewish Historical Society*, (1920), Vol. XXVII.

Jackson, Andrew. Andrew Jackson to General John Coffee, September 29, 1812, in Moser, Harold; Hoth, David R.; and Hoemann, George H., eds. *The Papers of Andrew Jackson, 1821-1824* (Knoxville: The Unversity of Tennessee Press), Vol. V.

James, Catalina Mason Myers Julian. *Biographical Sketches of the Bailey-Myers-Mason Families, 1776-1905* (Washington, D.C., 1908).

Jordan, John W., (ed.). *Colonial and Revolutionary Families of Pennsylvania: Genealogical and Personal Memoirs* (New York & Chicago, 1911), Vol. I. Reprinted (Baltimore, Md.: Clearfield Co. Inc., by Genealogical Publishing Co., Inc., 1994, 2004).

Journal of the Assembly of the State of New-York at their Fifty-First Session and Held at the Capitol in the City of Albany, 1828 (Albany: Printed by E. Croswell, Printer to the State, 1829. Vol. I and Vol. II).

Journal of the Common Council of the City of Schenectady From April 6, 1854, To April 4, 1855 (Keyser, Printer: 1855).

Judd, Peter Haring. *More Lasting Than Brass: A Thread of Family from Revolutionary to Industrial Connecticut* (Boston: Northeastern University Press, 2004).

Kabala, James S. " 'Theocrats' vs. 'Infidels': Marginalized Worldviews and Legislative Prayer in 1830s New York." *Journal of Church and State*, Volume 51, No. 1 (2009).

Keep, Austin Baxter. *History of the New York Society Library* (New York: De Vinne Press, 1908).

Kent, James. James Kent to Mordecai Myers, October 20, 1814. *Theodorus Bailey Myers Collection*, #2132. New York Public Library.

King, Marion. *Books and People: Five Decades of New York's Oldest Library* (New York: Macmillan, 1954).

Kohut, George Alexander. *Ezra Stiles and the Jews: Selected Passages From His Literary Diary Concerning Jews and Judaism With Critical and Explanatory Notes* (New York: Philip Lowen, 1902).

La Croix, Allison L. "To Gain the Whole World and Lose His Own Soul: Nineteenth-Century American Dueling As Public Law and Private Code". *Hofstra Law Review*, Vol. 33, 501-570 (2004).

Lamb, Martha J., and Harrison, Mrs. Burton. *History of the City of New York: Its Origin, Rise, and Progress* (New York: A.S. Barnes and Company, 1896), Vol. II.

Lear, Tobias. "Mr. Tobias Lear of the Dept. of War to Mr. R. Brent, relating to Cap M. Myers of 13th Inf, 1815." PID 365144; Call # P464 (courtesy, American Jewish Historical Society).

Leath, James Edward. *The Story of the Old House: Columbia County's "House of History"* (Kinderhook, N.Y.: The Columbia County Historical Society, 1947).

Lebeson, Anita Libman. *Recall to Life: The Jewish Woman in America* (New Jersey: Thomas Yoseloff, 1970).

Lengyel, Emil. *1,000 Years of Hungary* (New York: The John Day Company, 1958).

Lindegren, Alina Marie. *A History of the Land Bonus of the War of 1812* (University of Wisconsin, Unpublished Masters Thesis, 1922).

Litt, Paul; Williamson, Ronald F.; and Whitehorne, Joseph W.A. *Death at Snake Hill: Secrets from a War of 1812 Cemetery* (Ontario Heritage Foundation Local History Series No. 3. Toronto & Oxford: Dundurn Press, 1993).

Littell's Living Age. The Fifteenth Quarterly Volume of the Third Series. Volume 71, October, November, December, 1861 (Boston: Littell, Son and Company, 1861).

Lossing, Benson J. *Pictorial Field-Book of the War of 1812* (New York: Harper & Brothers, 1869).

Lossing, Benson John. *Notebooks and Journals.* Microfilm LS 1110-1128. The Huntington Library, San Marina, California.

Lowell, Robert. "91 Revere Street" in *Life Studies* (New York: Vintage Books, 1959).

Lowenthal, Larry. *Marinus Willett: Defender of the Northern Frontier* (Fleischmanns, NY: Purple Mountain Press, 2000), et passim.

Macartney, C.A. *Maria Theresa and the House of Austria* (Mystic, Connecticut: Lawrence Verry Inc., 1969).

Mackinnon, Neil. *This Unfriendly Soil: The Loyalists Experience in Nova Scotia 1783-1791* (Kingston and Montreal: McGill-Queen's University Press, 1986).

Maclay, Edgar Stanton; and Smith, Roy Campbell. *A History of the United States Navy, from 1775 to 1893* (New York: D. Appleton and Company, 1894), Vol. 1.

Mahon, John K. *The War of 1812* (Gainesville, Florida: The University Presses of Florida, 1972).

Malcolm, Richard. Major Richard Malcolm to Mordecai Myers, January 22, 1815. *Mordecai Myers Collection,* Clements Library, University of Michigan, Ann Arbor, Michigan.

Malcomson, Robert G. *Historical Dictionary of the War of 1812.* (Lanham, MD: Scarecrow Press, 2006).

Man, Albon. *No. 3 Accts Current, Constable, NY.* Courtesy of the family of Richard W. Floyd.

Mann, Allen B. *A History of Congregation Gates of Heaven.* Undated typed manuscript, circa 1938-1940. Courtesy Congregation Gates of Heaven.

Mann, James. *Medical Sketches of the Campaigns of 1812, 13, 14* (Dedham, MA: 1816), et passim.

Manumission Society, New York City, Minutes, Vol. IX, December 7, 1802 and January 18, 1803. MS, New York Historical Society.

Marcus, Jacob Rader. *American Jewry – Documents – Eighteenth Century* (Cincinnati, Ohio: The Hebrew Union College Press, 1959).

Marcus, Jacob Rader. *Memoirs of American Jews 1775-1865* (Philadelphia: The Jewish Publication Society of America, 1955. Volume I).*Museum Notes,* Toledo Museum of Art, 1957.

Marcus, Jacob Rader. *The Colonial American Jew, 1492-1776* (Detroit, Michigan: Wayne State University Press, 1970. Vol. II).

Marcus, Jacob R. *United States Jewry, 1776-1985* (Detroit: Wayne State University Press, 1989. Vol. I and II).

Martin, Hugh. Hugh R. Martin to Mordecai Myers, December 16, 1814. *Mordecai Myers Collection,* Clements Library, University of Michigan, Ann Arbor, Michigan.

Marton, Erno. "The Family Tree of Hungarian Jewry", in Braham, Randolph L., ed., *Hungarian Jewish Studies* (New York: World Federation of Hungarian Jews, 1966,1969. Vol. I).

Mason, George Champlin. *Reminiscenses of Newport* (Newport, R.I.: Charles E. Hammett, Jr., 1884).

Mather, Frederic Gregory. *The Refugees of 1776 from Long Island To Connecticut* (Albany,N.Y.: J.B.Lyon, Printers, 1913).

McGillivray, Robert; and MacGillivray, George B. *A History of the Clan MacGillivray* (privately printed, 1973).

McLean, John B. *A Brief History of the First Presbyterian Church of Oswego, N.Y. From Its Organization, November 21, 1816 to the Present* (Oswego: Times Book and Job Printers, 1890).

Mead, Spencer Percival. *History and Genealogy of the Mead Family of Fairfield County, Connecticut, Eastern New York, Western Vermont and Western Pennsylvania, from A.D. 1180 to 1900* (New York: The Knickerbocker Press, 1901).

Memorial of Daniel Lord (New York: D. Appleton and Company, 1869).

"Memorial Service for American-Jewish War Heroes 2006", *Veteran Corps of Artillery, State of New York* (retrieved from www.jag – consulting.com/VCA/PhotoGallery.php?album=3).

Merritt, Richard; Butler, Nancy; and Power, Michael, eds. *The Capital Years: Niagara-on-the-Lake, 1792-1796* (Toronto, ONT: Dundurn Press Limited, Co-published by the Niagara Historical Society, 1991).

Mika, Nick and Helma. *United Empire Loyalists: Pioneers of Upper Canada* (Belleville, Ontario: Mika Publishing Company, 1976).

Millard, James P. *Burlington, Vermont, during the War of 1812* (www.historiclakes.org/explore/burlington.htm).

Mines, John Flavel. *A Tour Around New York and My Summer Here, Being The Recreations of Mr. Felix Oldboy* (New York: Harper & Brothers, Publishers, 1893).

Minutes of Tammany Society or Columbian Order, 1791-1795, New York Historical Society.

M. Myers, *Military Pension File:* "Invalid". SC-27240: War of 1812. File No. 27, p. 240, Vol. 2, p. 293. National Archives, Washington, D.C.

Mooers, Benjamin. Benjamin Mooers to Mordecai Myers, September 4, 1812. Theodorus Bailey Myers Collection, #1852, New York Public Library.

Mooers, Benjamin. Benjamin Mooers to Mordecai Myers, September 5, 1812. Mordecai Myers Collection, The Clements Library, University of Michigan.

Moore, Christopher. *The Loyalists: Revolution, Exile, Settlement* (Toronto: Macmillan of Canada, 1984).

Mordecai Myers, *Letterbook, 1835-1841.* Courtesy of the late Dr. John S. Hoes.

Mordecai Myers, Reg. of Enl. – U.S.A. - Record Group No. 94, Military File. Book 405. Vol. 16, p. 161, #5353; Mordecai Myers, Capt. U.S. I. File No. 27, 240, Vol. 2, p. 293. National Archives, Washington, D.C.

Morgan, Willoughby. Willoughby Morgan to Mordecai Myers, March 21, 1815. *Mordecai Myers Collection,* Clements Library, University of Michigan, Ann Arbor, Michigan.

Morris, Ira K. *Memorial History of Staten Island* (New York: Memorial Publishing Company, 1898-1900), Vol. II.

Morris, John D. *Sword of the Border: Major General Jacob Jennings Brown, 1775-1828* (Ashland, Ohio: Kent State University Press, 2000).

Morton, Julius Stirling. *Illustrated History of Nebraska*. (Lincoln: Jacob North & Company, 1907), Vol. II.

Mowers, Kay. "Cantonment Farm," *Historic Albany Area No. 15*, December 17, 1941.

Mushkat, Jerome. *Tammany: The Evolution of a Political Machine 1789-1865* (Syracuse, New York: Syracuse University Press, 1971).

Myers, Benjamin. Benjamin Myers to Thomas Carleton, Petition, Grimross plot (sic), July 24, 1786, RS-108, Public Archives of New Brunswick, Fredericton, New Brunswick, Canada.

Myers, Gustavus. *The History of Tammany Hall* (New York: Dover Publications, Inc. 1971).

Myers, Mordecai. *Mordecai Myers. Reminiscenses 1780 to 1814 – Including Incidents In The War of 1812-14. Letters Pertaining To His Early Life* (Washington, D.C.: The Crane Company, 1900).

Myers, Mordecai. M. Myers to Genet, June 22, 1835. Mordecai Myers, *Letterbook, 1835-1841*. Courtesy of the late Dr. John S. Hoes.

Myers, Mordecai. [Mordecai] Myers to Mrs. (William) Hatch, July 5, 1840. Mordecai Myers, *Letterbook, 1835-1841*. Courtesy of the late Dr. John S. Hoes.

Myers, Mordecai. [Mordecai] Myers to Gideon Hawley, April 12, 1840. Mordecai Myers, *Letterbook, 1835-1841*. Courtesy of the late Dr. John S. Hoes.

Myers, Mordecai. Mordecai Myers to Charlotte M. Jackson, May 17, 1864. *Mordecai Myers Collection*, Clements Library, University of Michigan, Ann Arbor, Michigan. Vol. II.

Myers, Mordecai. Mordecai Myers to Theodorus Bailey Myers, October 26, 1860. *Mordecai Myers Collection*, Clements Library, University of Michigan, Ann Arbor, Michigan. Vol. II.

Myers, Mordecai. Mordecai Myers to Theodorus Bailey Myers, December 31, 1869. *Mordecai Myers Collection*, Clements Library, University of Michigan, Ann Arbor, Michigan. Vol. II.

Myers, Mordecai. Mordecai Myers to Col. P.P. Schuyler, June 7, 1812. *Mordecai Myers Collection*, Clements Library, University of Michigan, Ann Arbor, Michigan.

Myers, Mordecai. [Mordecai] Myers to Winfield Scott and Isaac Varian, November 16, 1841. Mordecai Myers, *Letterbook, 1835-1841*. Courtesy of the late Dr. John S. Hoes.

Myers, Mordecai. M. Myers to Henry Thomas, no date. Mordecai Myers, *Letterbook, 1835-1841*. Courtesy of the late Dr. John S. Hoes.

Myers, Rachel. Rachel Myers to Governor Carleton: Parr, May 16, 1785. Form for Land Grant Application Renewal, Grimross. May 16, 17, 1785.

Myers, Rachel. Rachel Myers to Thomas Carleton, Petition, Gagetown, June 15, 1786. RS-108, Public Archives of New Brunswick, Fredericton, New Brunswick, Canada.

Myers, Rachel. Rachel Myers to Thomas Carleton. Petition, August 29, 1786, Gage town. RS-108, Public Archives of New Brunswick, Fredericton, New Brunswick, Canada.

Myers, Rachel. Rachel Myers to Sir Henry Clinton. New York, April 3, 1781. Photocopy courtesy of the College of William and Mary: The Colonial Williamsburg Foundation.

Myers, Rachel. Rachel Myers to Samuel Dickenson (sic)....#1....1st Survey on Grimross Neck 1 May 1787 *Queens County Land Registers to 1866.* L.R.I.S. Office, Saint John, New Brunswick, Canada.

Myers, Rachel. Rachel Myers to *[probably the local land agent James Peters]* at Grimross Creek, May 30, 1785. RS-108, Public Archives of New Brunswick, Fredericton, New Brunswick, Canada..

National Archives of Canada. Vol. 1851, Reel C-3873; Vol. 1856, Reel C-4216; Vol. 1855, Reel C-3874; Vol. 1855, Reel C-3874.

New York City Directory (1797 and 1801-1810 inclusive).

New York City, Society of Tammany or Columbian Order, Constitution and Roll of Members, 1789-1816.

New York Mayors Court, Calendar of Cases. Reels IV, VI, VII, VIII and IX. Courtesy of the American Jewish Historical Society.

New York Mayors Court, Calendar of Cases. Insolvency Assignments – Insolvent Debtors. Reel III. Courtesy of the American Jewish Historical Society.

Niagara Falls: A Chronicle of Our Early Settlers – A History, 1600-1900. http://www.niagarafrontier.com/work.html.

Nolosco, Marynita Anderson. "Physician Heal Thyself: Medical Practitioners of Eighteenth Century New York". *American University Studies, Series IX, History: Vol. 170* (New York: Peter Lang Publishing, Inc. 2004).

Oberly, James W. "Gray-Haired Lobbyists: War of 1812 Veterans and the Politics of Bounty Land Grants." *Journal of the Early Republic,* Vol. 5, No.1 (Spring, 1985).

Paige, Lucius Robinson. *History of Hardwick, Massachusetts. With A Genealogical Register* (Boston: Houghton, Mifflin and Company, 1883).

Parton, James. *Life and Times of Aaron Burr* (New York: Mason Brothers, 1858).

Patai, Raphael. *The Jews of Hungary: History, Culture, Psychology* (Detroit: Wayne State University Press, 1996).

Peterson, Clarence Stewart. *Known Military Dead During the War of 1812* (Baltimore, Maryland: April 1955). Rept. Genealogical Publishing Co. Inc., 1991.

Polenberg, Richard. *The World of Benjamin Cardozo: Personal Values and the Judicial Process* (Cambridge, MA: Harvard University Press, 1997).

Pomerantz, Sidney I. *New York: An American City 1783-1803: A Study of Urban Life* (New York: Columbia University Press, 1938).

Pool, David De Sola. *Portraits Etched in Stone – Early Jewish Settlers, 1682-1831* (New York: Columbia University Press, 1952).

Pool, David and Tamar De Sola. *An Old Faith in the New World: Portrait of Shearith Israel 1654-1954* (New York: Columbia University Press, 1955).

Prime, Ebenezer. *Records of the First Church in Huntington, Long Island, 1723-1779* (Huntington, N.Y., 1899).

Proceedings of the Grand Chapter Royal Arch Masons of New York State at Its One Hundred and Seventeenth Annual Convention Held [at] Albany (New York: the Grand chapter, 1914).

Public Archives of New Brunswick, Fredericton, New Brunswick, Canada. "The Petition of Rachel Myers to His Excellency Thomas Carleton". Parr: January 24, 1785. RS-108.

Public Archives of Nova Scotia. Records Group 20. Series A, Vol. 12 (1784) #43.

Quimby, Robert S. *The U.S. Army in the War of 1812: An Operational and Command Study.* East Lansing, Michigan: Michigan State University Press, 1997. Vol. I).

Quincy, Eliza Susan. *Memoir of the Life of Eliza S.M. Quincy* (Boston, Massachusetts: John Wilson and Son, 1861).

Record in the Case of the United States of America Versus Fernando M. Arrendondo and Others. Supreme Court of the United States. January Term, 1831 (Washington: Printed By duff Green, 1831).

Reid, Robert W. *Washington Lodge No. 21, F. & A.M. and Some of Its Members* (New York: Washington Lodge Publishers, 1911).

Remini, Robert V. "The Albany Regency", New York History, 39 (Oct. 1958).

Rezneck, Samuel. "The Social History of an American Depression, 1837-1843", *The American Historical Review.* Vol. XL, No. 4, July, 1935.

Richards, Wilhelmina and Clifford, transcribers and editors. "The Recollections of Susan Man McCulloch", *Old Fort News* (Allen County-Fort Wayne Historical Society, Fort Wayne, Indiana. Vol. 44, No. 3, 1981).

Rottenberg, Dan. *Finding Our Fathers: A Guidebook to Jewish Genealogy* (New York: Random House, 1977).

Runciman, Steven. *A History of the Crusades* (London: The Folio Society, 1994. Vol. 1).

Russell, William. "The Landing of the United Empire Loyalists in New Brunswick", *Miscellaneous Research Papers, Manuscript Report Series, No. 216* (Ottawa: Parks Canada, 1975-77).

Ryerson, Egerton. *The Loyalists of America and Their Times from 1620 to 1816* (New York: Haskell House Publishers Ltd., 1970, Rpt. of 1880 ed., II).

Schappes, Morris U., ed. *A Documentary History of the Jews in the United States 1654-1875* (New York: Schocken Books, 1971).

Scharf, John Thomas. *History of Philadelphia, 1609-1884* (Philadelphia: L.H. Everts & Co., 1884), Vol. III.

Scharf, John Thomas. *History of Western Maryland, Being a History of Frederick, Montgomery, Carroll, Washington, Allegany, and Garrett Counties* (Philadelphia: L.H. Everts, 1882).

Schenectady City Directories, 1866-68 (Schenectady, New York).

Schenectady Surrogate Court Records, Records Room, Box No. 87.

Schlesinger, Jr., Arthur M. *The Age of Jackson* (New York: Little, Brown & Company, 1945).

Schreiber, A. *Jewish Inscriptions in Hungary From the 3rd Century to 1686* (Budapest and Leiden, 1983), pp. 314-318.

Scoville, Joseph Alfred. *The Old Merchants of New York City* (New York: Carleton, 1863).

Scudder Collection of Long Island Genealogical Records, (Huntington Historical Society).

Seaver, Frederick J. "History of Chateaugay, New York," in *Historical Sketches of Franklin County and Its Several Towns With Many Short Biographies* (Albany: J.B. Lyon Company, Printers, 1918), Chapter XI.

Seaver, Frederick J. "St. Regis", in *Historical Sketches of Franklin County and Its Several Towns With Many Short Biographies* (Albany: J.B. Lyon Company, Printers, 1918), Chapter XXIII.

Severance, Frank H. "The Earliest Map of Buffalo including an 1813 Burned Buffalo Map," in Frank Severance, ed., *The Picture Book of Earlier Buffalo.* Buffalo Historical Society, Vol. 16, 1912.

Sherwin, W.T. *Memoirs of the Life of Thomas Paine* (Ondon: R. Calile, 1819).

Shreve, Royal Ornan . *The Finished Scoundrel* (Indianapolis: The Bobbs-Merrill Company, 1933).

Skemp, Sheila. *A Social and Cultural History of Newport, Rhode Island 1720-1765* (Ann Arbor, Michigan: Xerox University Microfilms, 1974).

Snyder, Holly. "Reconstructing the Lives of Newport's Hidden Jews, 1740-1790", in *The Jews of Rhode Island.* Goodwin, George M. and Smith, Ellen, eds. (Waltham, Massachusetts: Brandeis University Press, 2004).

Skemp, Sheila. *A Social and Cultural History of Newport, Rhode Island, 1720-1765* (Ann Arbor, Michigan: Xerox University Microfilms, 1974), et passim.

Smith, R. Pearsall. *Gazetteer of the State of New York* (Syracuse, N.Y.: J.H. French, 1860).

Solis-Cohen, Myers. *The Hays Family in America* (1931, unpublished typed manuscript courtesy of the late Alva Middleton, the last descendant of Benjamin Myers).

Solis-Cohen, Salomon. "Note concerning David Hays and Reuben Etting, their Brothers, Patriots of the Revolution", *Publications of the American Jewish Historical Society,* Vol. II (1894).

Southwick, Solomon (pseudonym "Sherlock"). *A Layman's Apology, For The Appointment of Clerical Chaplains By the Legislature of The State of New York In A Series of Letters, Addressed To Thomas Hertell, 1833 – Signed* (Albany, New York: Hoffman & White, 1834).

Spafford, Horatio Gates. *A Gazetteer of the State of New York* (Interlaken, New York: Heart of the Lakes Publishing, 1981), Rpt. of 1825 edition.

Special and Local Laws Affecting Public Interests in the City of New York, In Force on January 1, 1880. George Bliss, Peter Olney, & William Whitney (compilers). (Albany: Charles Van Benthuysen & Sons, 1880. Vol. II).

Spooner, W.W. "The Van Rensselaer Family," *The American Historical Magazine,* Vol. 2, January 1907, No. 1.

Stern, Malcolm. "Myer Benjamin and his Descendants: A Study in Biographical Method", in *Rhode Island Jewish Historical Notes* (Providence, Rhode Island: Rhode Island Jewish Historical Association. Vol. 5, Number 2: 1968).

Stiles, Henry Reed. *Genealogies of the Stranahan, Josselyn, Fitch and Dow Families in North America* (Brooklyn, N.Y.: H.M. Gardner, Jr. Printers, 1868).

Stone, J. Gerald. *1895 Landmarks of Oswego County, NY* (September, 2004).

Stone, William Leete, (trans.). *Letters of Brunswick and Hessian Officers During the American Revolution* (Albany, N.Y. : Joel Munsell's Sons, Publishers, 1891).

Stryker, William S. "The New Jersey Volunteers – Loyalists – In The Revolutionary War", in *The Capture of the Block House at Toms River, New Jersey, March 24, 1782* (Trenton, New Jersey, 1883).

Sylvester, Nathaniel Bartlett. *History of Rensselaer County* (Philadelphia: Everts & Peck, 1880).

Taylor, Alan. *The Civil War of 1812, British Subjects, Irish Rebels & Indian Allies* (New York: Alfred A. Knopf, 2010).

The Constitution and Register of the membership of the General Society of the War of 1812 (Philadelphia: 1908).

The Military Society of the War of 1812 – Annals, Regulations, and Roster. Secretary and Adjutant's Office, March 12, 1895.

The New York City-Hall Recorder, For March, 1816 (New York: Clayton & Kingsland, 1816), Vol. I, No. 3.

Thompson, Zadock. *History of Vermont, National, Civil, and Statistical* (Burlington: Chauncy Goodrich, 1842).

Treman, Ebenezer Mack; and Poole, Murray L. *The History of the Treman, Termaine, Truman Family in America* (Ithaca, NY: Press of the Ithaca Democrat, 1901).

United States Naval Institute Proceedings. Volume 35 –Part 2 (Annapolis – Maryland: U.S. Naval Institute, 1909).

Unsigned. "Relating to the Military history of our Grandfather Major Myers", n.d. *Mordecai Myers Collection*, Clements Library, University of Michigan, Ann Arbor, Michigan. Vol. II.

U.S. Army Register of Enlistments 1798-1814, Entry #5748, National Archives, Washington, D.C.; Heitman, *HDRUSA*, Vol. I.

"USS Lady of the Lake," *Dictionary of American Naval Fighting Ships* (Naval History & Heritage Command: Department of the Navy – Naval Historical Center, Washington, D.C),www.history.navy.mil/danfs/l1/lady_of_the_lake-i.htm.

Van der Ar, Abraham Jakob. "Hellevoetsluis," in *Aardrijkskundig Woondenbock der Nederlanden, Vol. 9, J. Noorduyn,* (1884), pp. 389-393.

Vaughn, William Preston. *The Antimasonic Party in the United States, 1826-1843* (Lexington, Kentucky: University Press of Kentucky, 1983), et passim.

Veghazi, Istvan. "The Role of Jews in the Economic Life of Hungary" in *Hungarian Jewish Studies.* Braham, Randolph L., ed. (New York: World Federation of Hungarian Jews, 1966,1969. Vol. II).

Walton, E.P., ed. *Records of the Council of Safety and Governor and Council of the State of Vermont* (Monpelier: Steam Press of J. & J. M. Roland, 1873), Vol. I.

Whately, Harlan. "The History of the VCASNY (Veteran Corps of Artillery of the State of New York) and the Military Society of the War of 1812", Part I, February 19, 2008.

White, James Terry, ed. *National Cyclopedia of American Biography* (Clifton, NJ: J.T. White & Co., 1891), Vol. 13.

White, James. *Place-Names In The Thousand Islands, St Lawrence River* (Ottawa: Government Printing Bureau, 1910).

Willson, James R. *Prince Messiah's Claims To Dominion Over All Governments: And the Disregard of His Authority By the United States, In the Federal Constitution* (Albany, N.Y.: Packard, Hoffman and White, 1832).

Wilson, James Grant, ed. *The Memorial History of the City of New York, From Its First Settlement To the Year 1892.* (New York: New York History Company, 1893), Vol. III.

Wilson, James Grant; and Fiske, John. *Appleton's Cyclopedia of American Biography.* (New York: D. Appleton and Co., 1889).

Wright, Esther Clark. *The Loyalists of New Brunswick* (Fredericton, New Brunswick, Canada: 1955).

Wright, Silas. Silas Wright to [Mordecai] Myers, December 5, 1841. *Theodorus Bailey Myers Collection*, #2132. New York Public Library.

Yetwin, Neil B. "American-Jewish Loyalists: The Myers Family of New Brunswick", *The Loyalist Gazette* (Toronto: The United Empire Loyalists Association of Canada), Vol. XXVII, No. 2, Fall, 1990.

Yetwin, Neil B. "Dr. Albon Man: 'A Physician of Large Practice'." *Franklin County Historical Review* (Malone, New York: The Franklin County Historical Society & Museum), Vol. 26, 1989.

Young, Sue M. *A History of Amherst, N.Y.*, n.d., Chapters 2-3.

Zeichner, Oscar. "The Loyalist Problem in New York After the Revolution", *New York History* (New York Historical Society, XXI, 1940).

• *Mordecai Myers* — **Production Acknowledgements** •

Project Coordinator, Design, Layout: *Harry M. DeBan.*

Production Editor: *Harry M. DeBan.*

Proofing: *Henri-Michel duLac and N. B. Yetwin.*

Production Support: *Roland and Bonnie Nafus; David J. Bertuca; David Caldwell; Françoise and Sandrine duLac; Robert L. Emerson; Jerome P. Brubaker; Cynthia Liddell; Donald E. Graves; Keith G. Kozminski, Ph.D.; Dr. Laura G. Fischer; and, Tech. Sgt. Marty Abramson, USAF(Ret.).*

Graphics: *Neil B. Yetwin and Harry M. DeBan.* • **Cataloging:** *David J. Bertuca.*

Additional Research (Production): *Harry M. DeBan; Lt. Col. Richard V. Barbuto, USA (ret.).*

Pre-Press: *Rapid Service Engraving, Buffalo, NY (John Egloff and Joseph Castiglia).*

Neil B. Yetwin

Neil B. Yetwin has taught History, English and Psychology for 35 years, 32 of which have been at Schenectady (formerly Linton) High School in Schenectady, New York. He has published nearly 100 articles in a variety of journals and newspapers, including: *Yankee Magazine;* the *History Book Club Review; Columbia County History & Heritage; The Loyalist Gazette;* the *Emerson Society Papers;* the *Thoreau Society Bulletin; The Franklin County (NY) Historical Review; Black Diaspora; The Dukes County (MA) Intelligencer; New York Alive;* the *Albany Times Union;* the *Schenectady Gazette;* the *Schenectady County Historical Society Newsletter* and others, and has lectured extensively about local history at public schools, colleges, libraries, historical societies, places of worship and veterans organizations throughout New York's Capital District and beyond.

Neil was the 1989 recipient of the Louis B. Yavner Award of the New York State Regents for Excellence in Teaching the Holocaust and Civil Rights, and in 2005 was named "Outstanding Citizen of Schenectady" by the Schenectady City Council for his research and presentations about local history. His research on the Myers family experience as Loyalist refugees in Nova Scotia/New Brunswick, was cited in Sheldon J. and Judith C. Godfrey's book, *Search Out the Land: The Jews and the Growth of Equality in British Colonial America, 1740-1867* (Montreal & Kingston: McGill-Queen's University Press, 1995).

In addition to his interest in the War of 1812, Neil also enjoys the writings of Francis Parkman, Arthur Schopenhauer, Oswald Spengler, Carl Gustav Jung, and Henry David Thoreau. Neil's address at the July 2009 Thoreau Society Annual Gathering in Concord, MA, "'What demon possessed me that I behaved so well?': Thoreau's Jungian Shadow" was based on his article, "Thoreau, Jung, and the Collective Unconscious" (*Thoreau Society Bulletin,* Number 265, Winter 2009, pp. 4-7), and has been recognized as the first successful attempt to link the ideas of those two men. He was also the 2011 recipient of the "Preservation of History Award" for his contributions to the African-American Bural Ground Project at Schenectady's Vale Cemetery. In 2013, his published research about escaped slave Moses Viney helped establish Vale Cemetery as a historic site in the National Park Service *Underground Railroad Network to Freedom.*

Neil has been a presenter for the New York Council for the Humanities since 2010 with a lecture entitled, "Major Mordecai Myers: An American-Jewish Hero of the War of 1812." Yetwin has completed the first full-length biography of Mordecai Myers which he hopes to publish, and is currently writing a philosophy of history, *Birth, The First Act of War.* In his spare time he enjoys reading, walking in the woods, listening to music, and playing guitar in a band called *Running the River.* Neil and his wife reside in Schenectady, New York.

PRESERVING HISTORY AT OLD FORT NIAGARA

The Old Fort Niagara Association, Inc., is a 501(c)(3) Not-for-Profit organization, incorporated in 1927 to advance the preservation and restoration of the historic 1726 "Old" Fort Niagara within the United States Army's Fort Niagara Military Reservation, and to promote the history of New York State and the United States of America. Since completion of restoration efforts in 1934, it has continued to operate the site as a museum and educational institution, first, under license from the War Department, then, the State of New York when the land was transfered from Federal to State ownership. The Old Fort and the remainder of the Fort Niagara Military Reservation now comprise Fort Niagara State Park, except for U.S. Coast Guard Station Niagara — immediately below the west (river) wall of the old fortress — which remains an active duty Federal installation under the Department of Homeland Security.

Administration of Old Fort Niagara and several related structures is carried out by the Old Fort Niagara Association in cooperation with the New York State Office of Parks, Recreation and Historic Preservation. Ongoing programs and research cover the full span of Fort Niagara's and its predecessors' histories (1679-1963) under control of France, Great Britain, and the United States, and the military alliances and cultural interaction each maintained with the Native people of North America. The Old Fort is a registered National Historic Landmark, and was granted that status in 1962.

Publications are an extension of the Old Fort Niagara Association's educational purpose. Created in 1984, the Publications Committee has been charged with establishing and maintaining an ongoing program of works relevant to the history of Fort Niagara. This includes new titles, re-publication of older works, and production of other designated audio and visual materials.

Old Fort Niagara Association Publications Committee
Harry M. DeBan, Chairman and Publisher
David J. Bertuca • Craig O. Burt, III • R. Arthur Bowler, Ph.D.
David Caldwell • Lawrence Fortunato • Patricia Rice

Additional information about the Fort's publications, exhibits, and programs — as well as membership opportunities in the Old Fort Niagara Association — may be obtained from:

Old Fort Niagara Association
Fort Niagara State Park
P.O. Box 169
Youngstown, New York 14174-0169
http://www.oldfortniagara.org • ofn@oldfortniagara.org • (716)745-7611